Thanks and acknowledgements

The authors would like to thank Dr Colin Robertson, Dr Derek Bell, Dr Peter Williamson, Mrs P Tyler RM, Dr S. Glendinning, Dr E. Byrne, and Dr V. Hogg for help and advice on the medical content.

Thanks also to Nick Robinson and Lyn Strutt for their editorial input, and to the production department at Cambridge University Press.

The publishers would like to thank Nick Hillman and Bob Hubbard of Anglia Ruskin University, Judy Wick, and Dr Sue Nelson for reviewing the material prior to publication.

The authors and publishers are grateful to the following for permission to reproduce copyright material. While every effort has been made, it has not always been possible to identify the sources of all the material used, or to contact the copyright holders. If any omissions are brought to our notice, we will be happy to include the appropriate acknowledgements on reprinting.

The Scotsman for the extract on p. 22 taken from 'Nurses carry out surgery in effort to cut patient waiting lists' by Craig Brown, 28 April 2005;

Lothian NHS Board for the adapted text on p. 27 from the information booklet 'Going to hospital';

BMJ for the adapted material on p. 32, 'The foundation programme curriculum – Q and As' BMJ Careers, 9 April 2005, p. 71, 'Herpes zoster', p. 72, 'Trauma part 2: Non-ballistic trauma', M. D. Tsaloumas et al, 2 May 1998, p. 101 , 'Depression and unwanted first pregnancy, S. Schmiege and N. F. Russo, 3 December 2005, p. 101, 'Cannabis intoxication and fatal road crashes in France' B. Laumon et al, 10 December 2005, p. 119, 'Alcohol drinking in middle age and subsequent risk of mild cognitive impairment and dementia in old age', T. Antila et al, 4 September 2004, p. 119, 'Non-steroidal anti-inflammatory drugs and serious gastrointestinal adverse reactions', 1 March 1986, p. 123, 'Paternal age and schizophrenia – a population based cohort study', A. Sipos et al, 22 October 2004, p. 124, 'Age at retirement and long term survival of an industrial population', S. P. Tsai et al, 29 October 2005, p. 125, 'Incidence and prognosis of asthma and wheezing illness from early childhood to age 33', Dr P. Strachan et al, 11 May 1995, with permission from the BMJ Publishing Group;

W. B. Saunders for the adapted text on p. 38 and p. 45. Reprinted from *Clinical Medicine* 5th edition, by Kumar P. et al, pp. 418–419, © 2002 with permission from Elsevier.

Elsevier Health Science for the adapted text on p. 42 and p. 56 from *Davidson's Principles and Practice of Medicine*, 2005, on p. 49 from 100 Cases for Students of Medicines, eds M Gillmer et al, pp. 100–101 and p. 121 from *Epidemiology and Public Health Medicine*, 1999, by Vetter N. and Matthews I., and pp. 104–107 from *Macleod's Clinical Examination (Edinburgh)*, 1995, eds Munro J. & Edwards C., © Elsevier;

Hodder Education for the extract on p. 46 and photograph on p. 54 from *Chamberlain's Symptoms & Signs in Clinical Medicine*, John Wright, 1980, and on pp. 53, 56, 59, 60, 62, 64, 69, 128, *100 Cases in Clinical Medicine*, P. J. Rees et al, 2000;

The Times for the headlines on p. 57, 'Cases of HIV infection reach record high in the UK', 25 November, 2005, and p. 61. 'Smells can diagnose Alzheimer's', 14 December 2005, and for the adapted article and bar chart on p. 120 from 'Relentless rise of AIDS as HIV infections top 40m' by Sam Lister, 22 November 2005, © N I Syndication;

The Lancet for the adapted table on p. 119. Reprinted from *The Lancet*, p. 327, p. 489–492, Walt, R. et al, 'Rising frequency of ulcer perforation in elderly people in the UK, © 1986, with permission of Elsevier;

General Medical Council for extract on p. 98 taken from GMC Guidelines, www.gmc-uk.org;

Dr Colin Robertson for extracts on pp 126–127 from a talk on carbon monoxide poisoning.

The publishers are grateful to the following for permission to reproduce copyright photographs and material:

Key: l = left, c = centre, r = right, t = top, b = bottom

Alamy Images/©Image Source for p. 18, /©Stock Images for p. 75, /©Phototake Inc for p. 95; Corbis/©Royalty Free for p. 82, /©Zefa/Robert Llewellyn for p. 91, /©Royalty Free for p. 92(l), /©Royalty Free for p. 92(r), /©Royalty Free for p. 93; Custom Medical Stock Photo for p. 122; Downs Surgical for p. 90(b) (Artery Forceps), (Dissecting Forceps), (Scissors), (Retractor); Mediscan for pp. 54, 62, 63, 66, 71(b), 80, 81; National Psoriasis Foundation for p. 71(t); Pulse Picture Library for p. 76; Punchstock/©PhotoDisc Blue for p. 92(c); Rex for p. 90(b) (Scalpel); Science Photo Library/©James Stevenson for p. 24(r), /©Custom Medical Stock Photo for p. 90(t); John Walmsley for p. 23; Wellcome Trust Medical Photographic Library for p. 24(l).

Picture Research by Hilary Luckcock.

Contents

Introduction

Who is this book for?

Professional English in Use Medicine is designed to help those who want to read medical journals and textbooks more fluently. It will also help medical students preparing for an elective attachment in an English-speaking country, and medical professionals preparing to work in English or to take part in conferences conducted in English. The level of the book is intermediate to upper-intermediate. The model used is British English.

This book assumes you know, or are in the process of learning from your medical course, medical terms derived from Greek and Latin (such as *dyspnoea* and *uterus*). Our focus is on the ordinary English equivalents of those terms (such as *breathlessness* and *womb*), and on English words which are given a special meaning in medicine (such as *guarding* and *clubbing*).

This book is not intended to teach you about medicine although we have been careful to ensure that all the medical content is accurate. We have used a number of authentic sources including textbooks, reference works and common medical forms. We have also drawn on a corpus of Medical English developed by the Institute of Applied Language Studies at the University of Edinburgh.

You can use the book on your own for self-study, or with a teacher in the classroom, one-to-one or in groups.

How is the book organized?

The book has 60 two-page units. The first 46 are thematic, covering medical topics from **Health and illness** to **Research studies**. The remaining 14 units cover communication skills such as **Taking a history** and **Conference presentations**.

The left-hand page of each unit explains new words and expressions in contexts which make their meaning clear, and the right-hand page allows you to check and develop your understanding of the new language and how it is used, through a series of exercises.

There are six **appendices**, providing illustrations of parts of the body and types of medication, a useful list of medical abbreviations, and examples of verbs for instructions, lay terms and definitions that can be used when speaking to patients.

There is an **answer key** at the back of the book. Most of the exercises have questions with only one correct answer. But some of the exercises, including the **Over to you** activities at the end of each unit (see opposite), are designed for writing and/or discussion about yourself and your own experience.

There is also an **index**, which lists all the new words and expressions presented in the book and gives the unit numbers where they appear. It also indicates how the terms are pronounced.

The left-hand page

This page presents the key vocabulary for each theme or skills area. The language is introduced in a series of short texts, diagrams and tables, with each section indicated by a letter – usually A, B and C – and a clear title.

In addition to explanations of vocabulary, this page includes information about typical collocations (word combinations). In some cases, reference is made to websites where you can find further information on the text topic.

There are also **notes** on language points, for example where a particular grammatical form is associated with a word, or where the same word may have different uses.

The right-hand page

The exercises on the right-hand page give practice in using the words and expressions presented on the left-hand page. Some units contain diagrams or tables to complete; in others you may be asked to complete case notes or dialogues.

'Over to you' sections

An important feature of *Professional English in Use Medicine* is the **Over to you** section at the end of each unit. These activities provide you with the opportunity to use the new terms in relation to your own work and studies, or to express your own opinions.

Self-study learners can do this section as a written activity.

In the classroom, the **Over to you** sections can be used as a basis for discussion with the whole class, or in small groups. Learners can follow up by using the **Over to you** section as a written activity, for example as homework.

How to use the book for self-study

Check the **contents** page for the theme or communication skills area of interest to you. Read through the texts on the left-hand page. If you meet words which you consider important and which are not explained in the text, look at the index to see if they are explained in another unit. Do the exercises on the right-hand page and check your answers in the key. If you find you have made mistakes, go back to the left-hand page and read through the texts again. Some learners find it useful to keep a vocabulary notebook or vocabulary cards with the meaning of the new terms, common collocations and a sentence to show how they are used.

How to use the book in a classroom

Teachers can use this book to supplement more general course books. For most exercises, learners can compare answers together before they check the key. If they disagree or fail to find the right answer, teachers can provide feedback to help them find the correct answer. Some of the **Over to you** activities are suitable for role play.

We hope you enjoy using this book.

Professional English in Use

Professional English in Use Medicine is part of a new series of **Professional English in Use** titles from Cambridge University Press. These books offer vocabulary reference and practice for specialist areas of professional English. Have you seen some of the other titles available in the series?

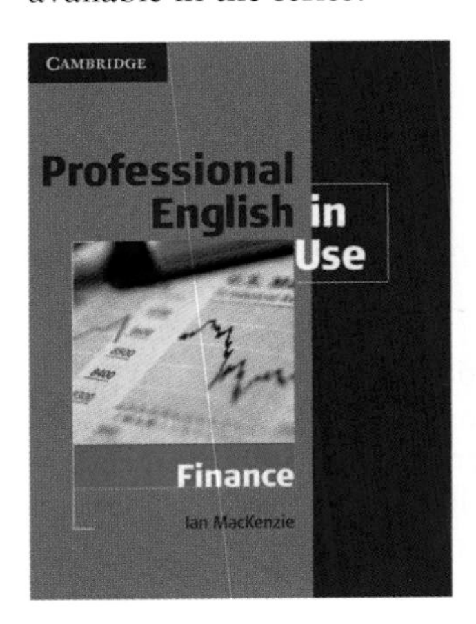

1 Health and illness

A Asking about health

Health is the state of the body. When doctors want to know about a patient's usual health, they ask questions such as:

> What is your **general health** like?

> How's your health, generally?

If you are **in good health,** you are **well** and have no **illness** (disease). If you are **healthy** you are normally well and can resist illness. If you are **fit,** you are well and strong.

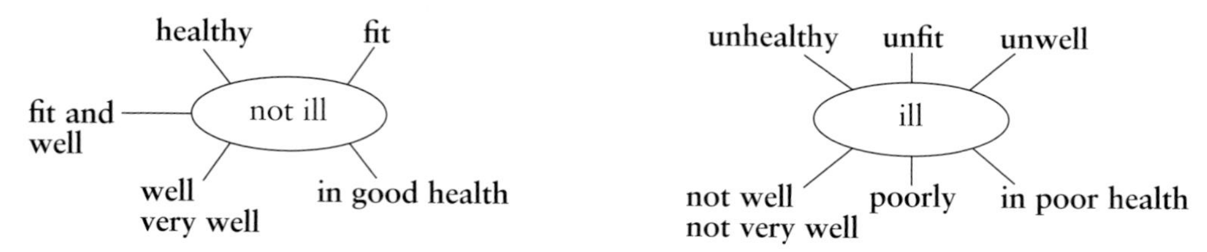

B Sickness

Sickness has a similar meaning to illness. It is also used in the names of a few specific diseases, for example **sleeping sickness** and **travel sickness.** Patients also talk about sickness when they mean nausea and vomiting.

Patient says	Possible meanings
I was sick this morning.	I was ill this morning. I felt unwell this morning. I vomited this morning.
I feel sick.	I feel ill. I feel unwell. I am nauseous. I feel the need to vomit.

The combination **sickness and diarrhoea** means vomiting and diarrhoea.

C Recovery

When patients return to normal health after illness, they have **recovered.** We can also say:

The patient	**made a**	**good** **full** **complete**	**recovery.**

If a patient's health is in the process of returning to normal, the patient is **improving.** The opposite is **deteriorating.** We can also say that the patient's condition **improved** or **deteriorated.**

In speech, we often use the verb **get** to talk about change:

get	**over** (an illness)	= to recover
	better	= to improve
	worse	= to deteriorate

If a patient is better, but then gets worse again, the patient has **relapsed.** Another word for **improvement,** especially in recurring conditions such as cancer, is **remission.**

> He **got over** the illness very quickly.

> Two years later she remains **in complete remission.**

1.1 Complete the table with words from A and B opposite. The first one has been done for you.

Noun	Adjective
fitness	fit
health	
illness	
sickness	

1.2 Make word combinations using a word from each box. Look at B and C opposite to help you.

complete feel get poor travel	sickness health remission sick over

1.3 Complete the conversation. Look at B opposite to help you.

Doctor: How are you feeling today?
Patient: Not very (1) ....
Doctor: How long have you been feeling (2) .... ?
Patient: About a week.
Doctor: What is your (3) .... like normally?
Patient: Very good. I'm usually quite (4) .... and (5) ....
Doctor: What is the problem now?
Patient: It's my stomach.
Doctor: Do you feel (6) .... ?
Patient: Yes.
Doctor: Have you actually been (7) .... ?
Patient: No.
Doctor: Have you had any serious (8) .... in the past?
Patient: No, none at all.

1.4 Choose the correct word to complete each sentence. Look at B and C opposite to help you.

1 Her condition .... (deteriorated/improved) and she died.
2 He .... (relapsed/recovered) and was allowed to go home from hospital.
3 The cause of sleeping .... (illness/sickness) was discovered in 1901.
4 The patient made a full .... (remission/recovery).
5 I have been in .... (poor/good) health for months and feel very fit.
6 It was a month before I .... (got over / got better) the illness.
7 He seems to be rather .... (unhealthy/unwell) – his diet is bad and he never exercises.

Over to you

What advice do you give people for keeping fit and well?

2 Parts of the body 1

A Parts of the body

Most external parts of the body have ordinary English names as well as anatomical names. Doctors normally use the English names, even when talking to each other. There are a few exceptions where doctors use the anatomical name; these are shown in brackets below.

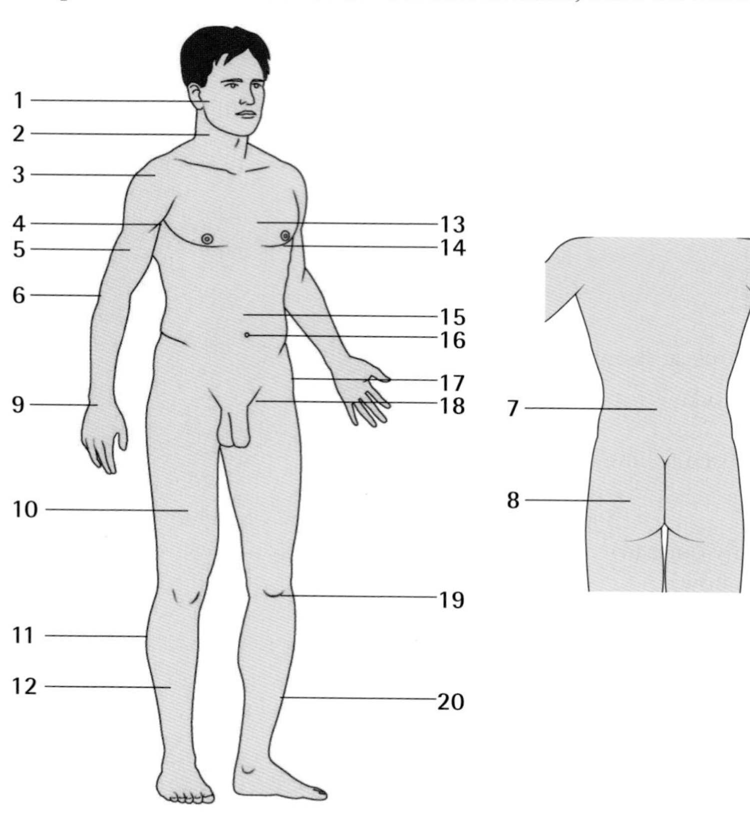

1 **jaw** (mandible)
2 **neck**
3 **shoulder**
4 **armpit** (axilla)
5 **upper arm**
6 **elbow**
7 **back**
8 **buttock**
9 **wrist**
10 **thigh**
11 **calf**
12 **leg**
13 **chest** (thorax)
14 **breast**
15 **stomach, tummy** (abdomen)
16 **navel** (umbilicus)
17 **hip**
18 **groin** (inguinal region)
19 **knee** (patella = **kneecap**)
20 **shin**

Limb means arm (**upper limb**) or leg (**lower limb**). The **trunk** is the body excluding the head and limbs.

For a more detailed diagram showing parts of the body, see Appendix I on page 130.

B Referring to parts of the body

When patients speak about their problem they often refer to a part of the body:

I'm **having trouble with my**	hip. shoulder. knee.

The doctor often needs to ask about a part of the body:

Do you get any **pain in**	the your	chest? stomach? back?

C Describing radiation of pain

A patient is telling the doctor about his back pain and the parts of the body it radiates to.

It **starts in** the back. Then it seems to **go into** the right buttock and **down** the back of the right thigh **to** the knee.

2.1 Write the ordinary English words for the corresponding anatomical terms in the table using your medical knowledge. Look at A opposite to help you.

Anatomical term	Common word
abdomen	
axilla	
carpus	
coxa	
cubitus	
mamma	
nates	
patella	

2.2 Complete the sentences using ordinary English words. Look at A and C opposite to help you.

a A male patient describing angina pectoris:

> It's like a tightness across my (1) .... , and it goes up (2) .... my (3) .... and into my left (4) .... and (5) .... the left (6) ....

b A male patient describing renal colic:

> It starts (1) .... the loin and goes into the (2) .... and (3) .... into the testicle.

2.3 Complete the sentences. Look at A opposite to help you.

Anatomical term	Patient's statement
1 inguinal swelling	I've got a lump in the ....
2 abdominal pain	My little boy's got a .... ache.
3 periumbilical rash	I've got some spots around my ....
4 thoracic pain	I've got a pain in the middle of the ....
5 enlarged axillary node	There's a painful swelling in my ....
6 mandibular pain	I've got a pain in my ....

2.4 Complete the table with words from the box. The first one has been done for you.

abdomen	elbow	loin	wrist	thigh
knee	chest	arm	leg	finger

Trunk	Upper limb	Lower limb
abdomen		

Over to you

Make a list of the words from A opposite that you find it hard to remember or that you need most often. Try to learn at least one of them every day.

3 Parts of the body 2

A The abdomen

The main **organs** of the body have ordinary English names and doctors use these words. But when an adjective is needed they often use an anatomical word. For example, we can say **disease of the liver** or **hepatic disease.** Some abdominal organs, for example the pancreas, have no ordinary name.

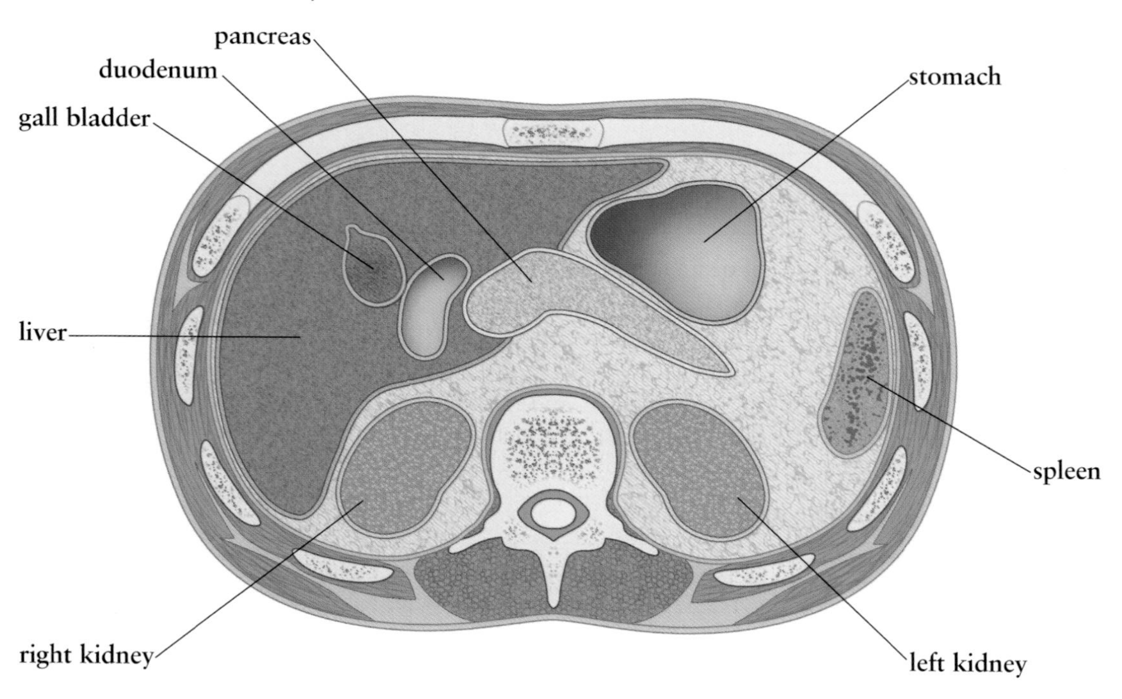

A cross-section of the abdomen, viewed from below

When doctors talk about the main parts of the digestive system, they use the words **bowel** or **intestine:** the **small intestine** or the **small bowel,** the **large intestine** or the **large bowel.** When speaking to patients, doctors may refer to the anus and rectum as the **back passage.**

B The chest

The chest (thorax) contains the organs of respiration and the **heart.** The main parts of the respiratory system are the **airways** and the **lungs.** The left lung is divided into two **lobes,** and the right into three. The airways consist of the larynx, the trachea (or **windpipe**), the right and left bronchus, and the **bronchioles.** The chest is separated from the abdomen by the **diaphragm.**

C The pelvis

A doctor is explaining the function of the **bladder** to a patient.

> The bladder is situated in the pelvis, as you know, and it is connected to each **kidney** by a long tube called the **ureter** – one on each side. The ureters carry the urine from the kidneys to the bladder, where it is stored until you decide to **empty your bladder.** When that happens, the urine passes down another tube, called the **urethra,** to the outside.

3.1 Label the diagram using words from the box. Look at B opposite to help you.

diaphragm	lobes	windpipe	heart
lung	airways	bronchioles	

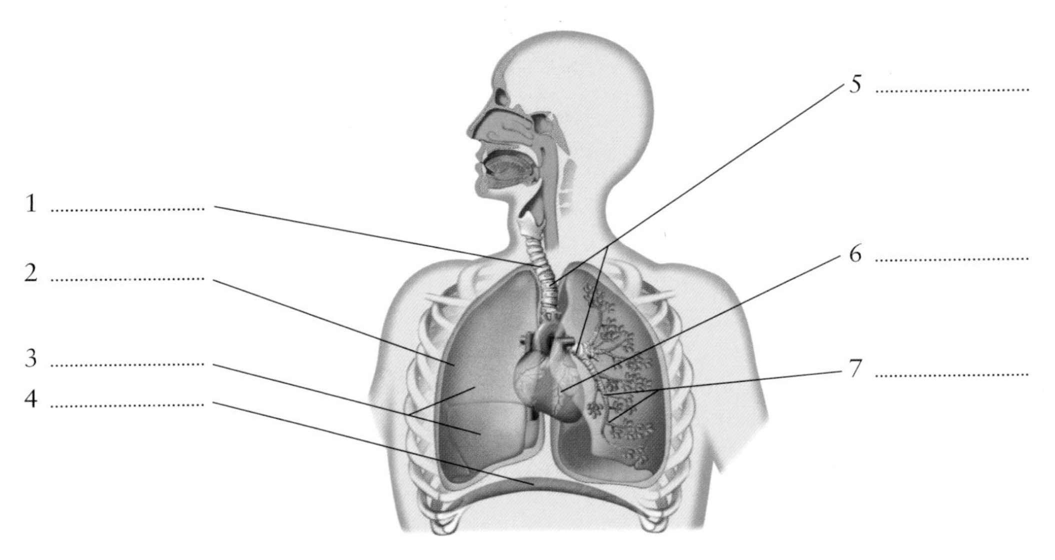

3.2 Match the conditions (1–8) with the organs affected (a–h), using your medical knowledge.

1 hepatitis	a bladder
2 pneumonia	b gall bladder
3 nephritis	c heart
4 gastric ulcer	d kidney
5 cystitis	e liver
6 angina pectoris	f lung
7 cholecystitis	g stomach
8 ulcerative colitis	h large bowel

3.3 Complete the textbook extract. Look at A and C opposite to help you.

Examination of the abdomen

To examine the patient for enlarged abdominal (1) .... , first feel for the (2) .... and the (3) .... on the right side. To do this, ask the patient to take a deep breath, while pressing with the fingers upwards and inwards. Next, feel for the right (4) .... and then cross over to the other side for the left (5) .... Still on the left side, palpate for an enlarged (6) .... Finally, moving to the lower abdomen, feel for the (8) .... , which is only felt if it is full.

Over to you

Many patients do not know the location or function of the spleen or the pancreas. How would you explain them to a patient, in English?

4 Functions of the body

A Eating

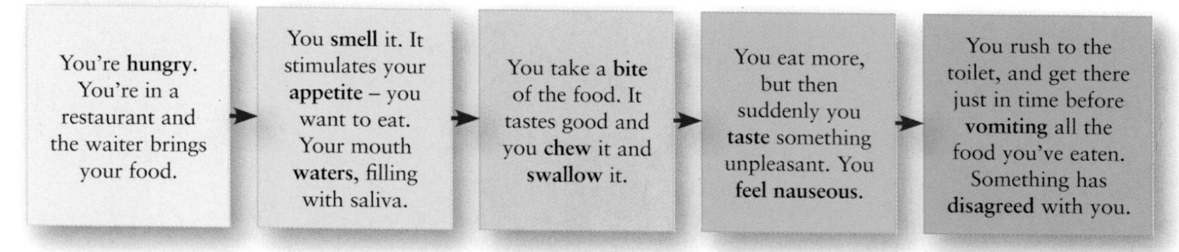

B The five senses

In addition to **smell** and **taste**, the senses include **sight** (or vision), **hearing**, and **touch** (also called **sensation** or **feeling**). To ask about the senses, doctors use the questions:

What is your	sight hearing **sense of** smell sense of taste	like?
Is your		normal?

To ask about the sense of touch, doctors talk about **numbness** (loss of sensation):

Have you noticed any numbness (in your fingers or toes)?

C Other functions

Function	Verb	Noun
speaking	speak	**speech**
walking	walk	**gait**
breathing respiration	inhale / **breathe in / take a breath in** exhale / **breathe out**	**breath**
urination micturition	urinate micturate **pass urine / pass water**	urine
defecation	defecate **pass faeces / pass stools**	faeces **stools**
menstruation	menstruate **have a period**	**(menstrual) period** **(monthly) period**

When taking a history, doctors can ask:

Do you have any	**trouble** **difficulty** **problems**	walking? breathing? passing urine? with your speech?
	pain	when you breathe in?

When auscultating a patient's lungs, the doctor tells the patient:

Take a deep breath in, hold your breath, then breathe out completely.

D Less common functions

There are some things we do less often. When we are hot, we **sweat**. When we are nervous, we **shake**. When we are sad, we **cry**.

Doctors can ask:

Do you	sweat shake	**more than usual?**

4.1 Match the symptoms (1–5) to the questions (a–e), using your medical knowledge.

1 dysuria
2 dysphagia
3 diplopia
4 dysphasia
5 dyspnoea

a What is your breathing like?
b Do you have any pain when you pass water?
c Do you have any difficulty with your speech?
d Do you have any trouble swallowing?
e Is your vision normal?

4.2 Patients are describing symptoms of the conditions shown in brackets. Complete the sentences. Look at C and D opposite to help you.

1 I've got pain and .... in both feet. (peripheral neuropathy)
2 I'm having difficulty .... solid food. (oesophageal stricture)
3 I have a lot of problems .... (prostatic hypertrophy)
4 I've been .... more than usual, even when it's not hot. (hyperthyroidism)
5 I've noticed that my hands .... when I'm not using them. (Parkinsonism)
6 I have trouble .... when I climb the stairs. (left heart failure)

4.3 Complete the sentences. Look at A, B, C and D opposite to help you.

1 When I eat solid food, I have to .... (bite/chew) it for a long time before I can .... (swallow/eat) it.

2 Do you have any pain when you .... (pass/have) stools?

3 I have no .... (taste/appetite) and I've lost five kilos in the last few weeks.

4 When did you last .... (have/pass) a period?

5 The garden is full of flowers, but my .... (sense/sensation) of smell has disappeared and I can't enjoy the perfume.

6 Take a deep .... (breathe/breath) in.

Over to you

You think a patient may have diabetes. Think of five questions you can ask the patient to investigate further. Try to use the question types presented in this unit.

5 Medical practitioners 1

A Practitioners

In Britain, doctors, also known as **medical practitioners**, must be **qualified**: have a university degree in medicine. They must also be **registered** – included in the General Medical Council's list, or **register** – in order to **practise** (see Unit 13). A doctor who treats patients, as opposed to one who only does research, is called a **clinician.** A doctor who provides primary care for patients is known as a **general practitioner (GP)**, or family doctor. GPs usually work in a **group practice.** Larger group practices work in a building called a **health centre.**

Note: In British English, the verb is spelt **practise** and the noun is spelt **practice**.

B Specialties

Specialist doctors, for example paediatricians, generally work in hospitals. However, those who work outside the NHS, providing **private health care**, may have **consulting rooms** outside a hospital – for example in the famous Harley Street in London.

The two main branches of medicine are **surgery** and **internal medicine**, and the doctors who practise these branches are called **surgeons** and **physicians**, respectively. In Britain, male surgeons are addressed as Mr and females as Ms – so Dr Smith is a physician, and Ms Smith is a surgeon.

<table>
<tr><td>A cardiologist</td><td rowspan="3">specializes in
is a specialist in</td><td>diseases of the heart and circulation, or cardiology.</td></tr>
<tr><td>A geriatrician</td><td>diseases of elderly patients, or geriatrics.</td></tr>
<tr><td>An anaesthetist</td><td>anaesthetics.</td></tr>
</table>

Note: Names of specialties usually end in **-ology**; names of specialists usually end in **-ologist**. If the name of a specialty ends in **-ics**, the name of the specialist ends in **-ician**. There are some exceptions, e.g. **anaesthetics** and **anaesthetist**.

C Choosing a specialty

Jill Mathews has just graduated from medical school and is talking about her future.

'I haven't decided what to **specialize in** yet. I need more experience before I decide, but I'm quite attracted to the idea of paediatrics because I like **working with** children. I'd certainly prefer to work with children than, say, elderly patients – so I don't fancy geriatrics. I was never very **interested in** detailed anatomy, so the **surgical specialties** like **neurosurgery** don't really appeal. You have to be **good with** your hands, which I don't think is a problem for me – I've **assisted at** operations several times, and I've even done some minor ops by myself – but surgeons have to be able to do the same thing again and again without getting bored, like tying off cut arteries and so on. I don't think that would be a problem for me, but they need to make decisions fast and I'm not too **good at** that. I like to have time to think, which means surgery's probably not right for me.'

Note: The collocation **good with** is followed by a noun – *He's good with children.* The collocation **good at** is followed by the -ing form (gerund) of a verb, or by a noun – *She's good at explaining procedures. She's good at explanations.*

5.1 Write sentences to describe the work of the specialist in each branch of medicine. Look at B opposite to help you.

1 dermatology — *A dermatologist specializes in diseases of the skin.*
2 rheumatology
3 traumatology
4 paediatrics
5 obstetrics

5.2 Complete the table with words from A, B and C opposite and related forms. Put a stress mark in front of the stressed syllable in each word. The first one has been done for you.

Verb	Noun (person)	Noun (activity or thing)
'specialize		
practise		
consult		
assist		
graduate		
qualify		

5.3 Find prepositions in C opposite that can be used to make word combinations with the words in the box. Then use the correct forms of the words to complete the sentences.

good	interested	specialize	work

1 A pathologist .... diagnosing disease through examining cells and tissue.
2 A paediatrician must enjoy .... children.
3 An oncologist is .... the diagnosis and treatment of cancer.
4 A psychiatrist must be .... counselling.
5 A neurosurgeon must be .... her hands.

5.4 Make word combinations using a word from each box. Two words can be used twice. Look at A, B and C opposite to help you.

consulting	centre
general	practice
group	specialties
health	medicine
internal	practitioner
surgical	rooms

Over to you

Re-read what Dr Jill Mathews says about surgeons in Section C. Make a list of the qualities she thinks are needed to be a good surgeon. Then make a similar list of qualities for another specialty.

If you are a student, which branch of medicine do you think you have the qualities for? If you have already completed your training, why did you choose your particular branch of medicine?

6 Medical practitioners 2

A Hospital staff

The people who work in any type of workplace, including hospitals, are called the **staff**. The **medical staff** in a British hospital belong to one of four main groups:

- A **pre-registration house officer (PRHO)**, or **house officer**, is a newly graduated doctor in the first year of postgraduate **training**. After a year, he or she becomes a registered medical practitioner. In the current system of training, the **Foundation Programme**, the name for these junior doctors is **Foundation Year 1 doctor** (**FY1**). (See Unit 12)
- A **senior house officer** (**SHO**) is in the second year of postgraduate training. The title is now **Foundation Year 2 doctor** (**FY2**), but the old terms senior house officer and SHO are still used.
- A **specialist registrar** (**SpR**) is a doctor who has completed the Foundation Programme, and is training in one of the medical specialties. There are also some **non-training registrars** – doctors who have completed their training but do not wish to specialize yet.
- A **consultant** is a fully qualified specialist. There may also be some **associate specialists** – senior doctors who do not wish to become consultants. In addition, there is at least one **medical** (or **clinical**) **director**, who is responsible for all of the medical staff.

B Medical teams

Consultant physicians and surgeons are responsible for a specific number of patients in the hospital. Each consultant has a **team** of junior doctors to help care for those patients. In many hospitals, there are **multidisciplinary teams** which consist not only of doctors but also of physiotherapists and other allied health professionals (see Unit 8).

When patients enter – or **are admitted to** – hospital, they are usually seen first by one of the junior doctors on the **ward** where they will receive treatment and care. The junior doctor **clerks them** – **takes their medical history** (see Units 47–49) – and examines them. Some time later, the registrar also sees the patients, and may order **investigations** or **tests**, for example X-rays or an ECG, make a provisional **diagnosis**, and begin treatment. The consultant usually sees the **new admissions** – people who have recently been admitted to the ward – for the first time on one of the regular **ward rounds**, when the management of the patients is discussed with the registrar. Consultants also decide when a patient is ready to be **discharged** (sent home). On the ward round, the consultant is accompanied by the team and a nurse, and they visit all the patients in the consultant's care.

C Shifts

Junior doctors now normally work in **shifts**, which means they normally work for eight hours every day, for example 7 am to 3 pm, and are then free until 7 am the next day. After a week they change to a different shift, for example 3 pm to 11 pm or 11 pm to 7 am. The alternative system is to work from 9 am to 5 pm every day and to take turns to be **on call** – available to return to the hospital if necessary – from 5 pm to 9 am the next day. Days on call are set out in a **rota**, or list of names and times. Doctors on call carry a **radio pager**, or **bleeper**, a device which makes a noise when someone is trying to contact them.

6.1 Make word combinations using a word from each box. Look at A, B and C opposite to help you.

associate	call
house	diagnosis
on	officer
provisional	pager
radio	round
ward	specialist

6.2 Match the descriptions (1–5) with the job titles (a–e). Look at A and B opposite to help you.

1 Dr Graham has been a paediatrician for eight years and is responsible for treating the children admitted to Ward 60.
2 Dr Stewart has just started the second year of her Foundation Programme.
3 Dr Singh has started his training as a surgeon.
4 Dr Phillips has just graduated and is working in a large hospital in Birmingham.
5 Dr Millar is in charge of the medical staff in the Birmingham hospital.

a specialist registrar
b medical director
c consultant
d SHO
e PRHO or house officer

6.3 Are the following statements true or false? Find reasons for your answers in A, B and C opposite.

1 A medical graduate becomes registered two years after graduation.
2 The system of training doctors in Britain is called the Foundation Programme.
3 The name senior house officer is no longer used in Britain.
4 The consultant is usually the first doctor to see new patients.
5 When working in shifts, all doctors take turns to be on call.

6.4 Complete the text of a PRHO describing her job. Look at A, B and C opposite to help you.

When I get to the ward, the first thing I do is talk to the house officer who was on duty during the last (1) .... , to find out if there have been any new (2) .... Then I generally see the charge nurse. He tells me if there is anything that needs to be done urgently, such as intravenous lines to put up or take down. Later in the morning, I (3) .... any new patients, which basically involves taking a history. On Tuesday and Friday morning the consultant does her ward (4) .... , and I have to make sure I'm completely up to date on her patients. After that, there are usually lots of things to do, like writing up request forms for blood (5) .... , and so on. In the afternoon, I have to prepare for any patients who are to be (6) .... the next day. They're usually happy to be going home! And then of course there are the lectures and tutorials in the (7) .... programme on Monday and Wednesday.

Over to you

How does the hospital training of doctors in your country differ from the British system? How would you explain it to a colleague from another country?

7 Nurses

A Nursing grades

Nurses working in a hospital have the following grades:

student nurse	a nurse who is still in training
staff nurse	a nurse who has completed the training course
charge nurse	a more experienced nurse who is **in charge of**, or responsible for, a ward or department
nurse manager	a nurse who is in charge of several wards

Note: The old term **sister** is still sometimes used for a female charge nurse. A female nurse manager may be called **matron**.

Dr James is talking to Sister Watkins.

B Support workers

The **clinical support worker**, who has done a short course and obtained basic qualifications, and the **nursing auxiliary**, who is usually unqualified, both assist nursing staff. There may also be **ward clerks**, whose duties include making sure patients' notes and information are up to date, and answering the telephone.

C Specialization

Like doctors, nurses can specialize:

- A **midwife** has specialized from the beginning by doing a course in midwifery, the management of pregnancy and childbirth.
- **District nurses** visit patients in their homes.
- **Health visitors** also work in the community, giving advice on the promotion of health and the prevention of illness.

D The nurse's role

The nurse's role has changed considerably in recent years. In addition to general **patient care, checking temperatures, pulse rates** and **blood pressures, changing dressings, giving injections** and **removing sutures,** nurses now do some of the things previously reserved for doctors, such as **prescribing drugs,** and **ordering laboratory tests.** More responsibility for nurses is planned, as the following article demonstrates.

Nurses carry out surgery in effort to cut patient waiting lists

Nurses in Scotland trained to **perform** minor surgery have entered the operating theatre for the first time in an effort to cut patient waiting times. Five nurses who have passed a new course at Glasgow Caledonian University are now qualified to **carry out** such **procedures** as the removal of small lesions, benign moles and cysts.

The Scotsman

The verbs **perform** and **carry out** are used with all types of procedures. They are often used in the passive form.

perform carry out	an examination an operation a procedure an experiment a test a biopsy

The procedure was	performed carried out	by a nurse.

7.1 Complete the sentences. Look at A, B and C opposite to help you.

1 Someone who specializes in delivering babies is a ....
2 Someone who is qualified to assist nurses is a ....
3 Someone who is not qualified but is able to assist nurses is a ....
4 A nurse who has qualified is a .... nurse.
5 A nurse who specializes in health promotion is a ....
6 A nurse who looks after a ward is a .... nurse.
7 A nurse who works in the community is a .... nurse.
8 Someone who answers the ward telephone is a ....

7.2 Make word combinations using a word or phrase from each box. Look at D opposite to help you.

carry out
change
check
give
remove

sutures
a procedure
an injection
a dressing
the temperature

7.3 Complete the sentences with the correct grammatical form of *perform*.

1 An isotope brain scan is painless and easy ....
2 Biopsy of the pancreas .... last March.
3 If the patient's condition deteriorates, a laparotomy should ....
4 If a diagnosis of meningitis is suspected a lumbar puncture must ....
5 Last year we .... a randomized, double blind group study.

7.4 Complete the sentences with the correct grammatical form of *carry out*.

1 I now intend .... a larger study.
2 Unfortunately few properly controlled trials .... so far.
3 A number of studies .... recently to look at this question.
4 A right hemicolectomy .... and the patient made a full recovery.
5 This procedure can .... in the emergency department.

Over to you

What kind of tasks do nurses carry out in your country? Are nurses' responsibilities increasing? What are the implications of this?

8 Allied health professionals

A Community health

The health of the community depends on a large number of people other than medical practitioners and nurses. These can be grouped under the heading of **allied health professionals**. They include the following:

- **Physiotherapists** (**physios**) help people to move by getting them to do exercises or by treating their body with heat or **massage** – treatment by manipulating muscles and joints with the hands. (See Unit 42)
- **Occupational therapists** (**OTs**) help people with a disability to perform tasks at home and at work. A **disability** is a physical or mental condition that makes it difficult to live normally, for example blindness or deafness.
- **Social workers** help people to solve their social problems – for example poor housing or unemployment – or family problems.
- **Chiropodists**, also know as **podiatrists**, treat conditions affecting the feet.

B Technicians

There are numerous **technicians** – people who work with scientific equipment – such as radiographers, who are known as **X-ray technicians. Ambulance technicians** work in the emergency medicine service. An ambulance technician with more advanced qualifications is called a **paramedic**.

C Prosthetists and orthotists

Prosthetists and **orthotists** provide care for anyone who needs an **artificial limb**, (a **prosthesis**), or a device to support or control part of the body (an **orthosis**). They also advise on **rehabilitation** – helping patients return to normal life and work after treatment.

Prosthetists provide **artificial replacements** for patients who have had an **amputation** or were born without a limb.

Orthotists provide a range of **splints** and other devices to aid movement, correct **deformity** from an abnormal development of part of the body, for example **club foot** (talipes), and **relieve pain**.

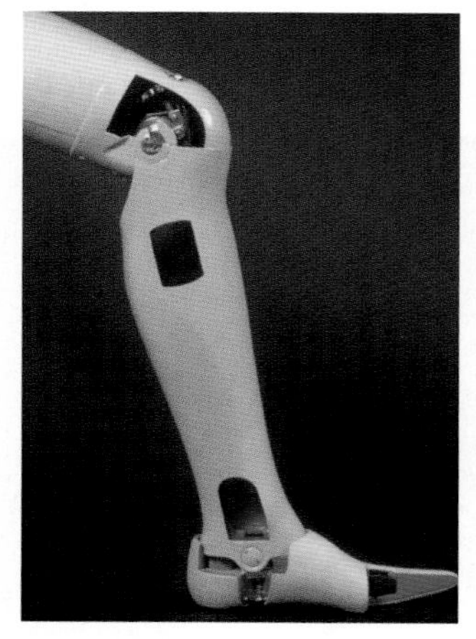

A prosthesis

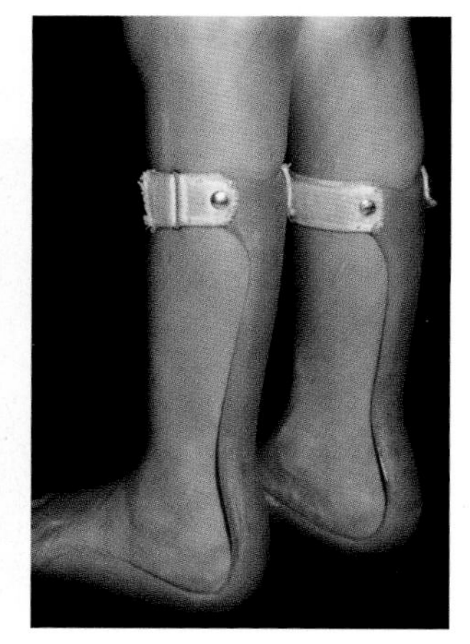

Splints

D Opticians

Opticians test **eyesight** and prescribe **glasses** – also know as **spectacles** – and **contact lenses**, when necessary. The examination includes measuring **intraocular pressure** – the pressure of fluid inside the eye – and examining the retina. If the optician suspects an eye disease, such as glaucoma, they refer the patient to their GP for treatment. The GP may then refer the patient to an **ophthalmologist**, a doctor who specializes in diseases of the eye.

8.1 Make word combinations using a word from each box. Look at A, B, C and D opposite to help you.

ambulance artificial club contact health intraocular occupational social	foot lens limb worker pressure technician professional therapist

8.2 Which allied health professionals could best help the following people? Look at A, B, C and D opposite to help you.

1 a young unmarried woman who has just had a baby
2 a woman who is having difficulty using her right arm following a fracture
3 someone who needs glasses
4 an elderly woman who has had a below knee amputation
5 a man whose wife has Alzheimer's disease
6 a man with a fungal infection of his feet

8.3 Complete the texts. Look at A and C opposite to help you.

A prosthetist works with patients of all ages as a member of a clinical team, based at a large hospital. The patients may need a prosthesis as the result of an accident, or (1) .... following a disease such as diabetes. Alternatively they may have been born without a (2) .... Orthotists work alongside doctors, nurses, physiotherapists and occupational (3) .... to give the people under their care the best possible (4) .... Their main aim is to enable the patient to lead a normal life at work and leisure.

An orthotist often works in a clinic as part of an outpatient service and also visits other centres to provide a service for people with special needs. They deal with people of all ages. For instance, children who have cerebral palsy may require (5) .... to help them walk and many older people need special shoes to correct (6) .... If damaged, any part of the human skeleton may require some form of orthosis. The orthosis may be needed to reposition the body or to (7) .... pain.

Over to you

Britain is introducing a new member to the healthcare team, called a medical care practitioner (MCP), similar to the physician assistant in the United States and other countries. The MCP will be able to carry out some of the functions of a medical practitioner, such as history-taking and examination, and diagnosis and treatment of certain illnesses, without having a medical degree. What are the advantages and disadvantages of this in your opinion?

9 Hospitals

A Introduction to a hospital

Jordi Pons is a fourth-year medical student from Barcelona. He has come to Britain on an elective attachment to the Royal Infirmary, Edinburgh. Dr Barron is introducing him to the hospital.

Dr Barron: The Royal Infirmary is the name of the **university hospital** for Edinburgh University. It is a **general hospital**, dealing with all types of patients and illnesses, except paediatrics. We have a **specialist hospital** for that in another part of Edinburgh, the Hospital for Sick Children. You can see some of the **departments** in our hospital on the sign. Of course, there are many others, for example the **Intensive Care Unit** (ICU), and the **Surgical High Dependency Unit** (HDU).

Jordi: What does '**outpatient**' mean?

Dr Barron: Outpatients are the people who come to hospital to **attend a clinic** or to **have tests** or **treatment** and then return home on the same day. **Inpatients** stay in the hospital for one or more days. The rooms where they stay are called **wards**. If a patient's treatment requires only one day, such as a simple operation, they can be admitted to the **day surgery unit**.

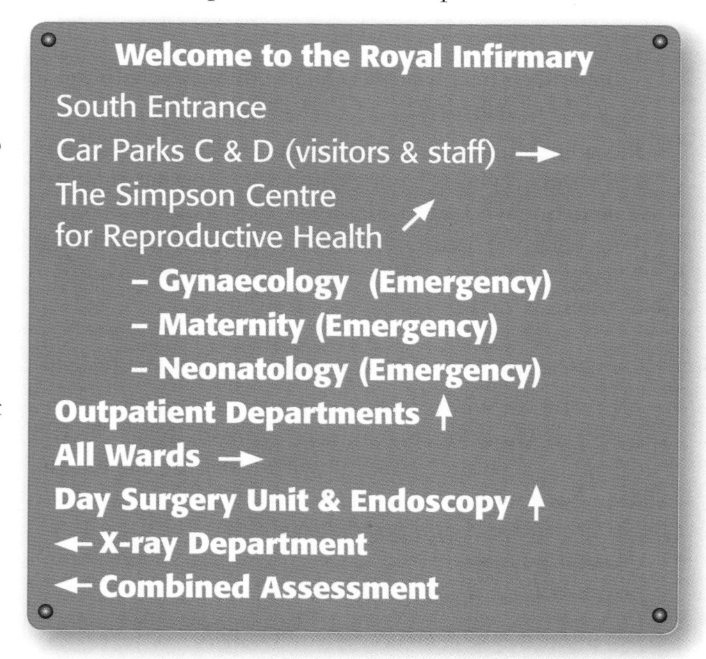

B Outpatients

Dr Barron: The **Accident and Emergency Department** (A&E), also called **Casualty**, is where patients who are **acutely ill** – with a sudden, serious condition – come for assessment and treatment. Outpatients who have an appointment to see a specialist go to a clinic in one of the **Outpatient Departments** (**OPDs**). They have usually been **referred to** the hospital by their GP, who writes a **referral letter** to the consultant explaining the patient's problem.

C Inpatients

Dr Barron: The inpatients in a hospital are **admitted** in one of three main ways. They may be seen in one of the outpatient clinics and admitted from there or, if there is a lot of demand for the treatment they need, as in the case of a hip replacement, they are **put on a waiting list** for admission. Alternatively, their GP may arrange the **admission** by telephone because they are acutely ill, for example with suspected myocardial infarction. Or they are seen in the A&E Department, where the doctor **on duty** – working at that time – arranges the admission. This would happen in the case of a patient with a fractured neck of femur, for example. Larger hospitals may have an **assessment unit** where patients can be admitted temporarily while their condition is **assessed**.

Jordi: Assessed?

Dr Barron: Yes – decisions are made about their condition, and what needs to be done to help them. After treatment is completed, the patient is **discharged** back to the GP's care.

9.1 Complete the table with words from A, B and C opposite. Put a stress mark in front of the stressed syllable in each word. The first one has been done for you.

Verb	Noun
ad'mit	
assess	
discharge	
operate	
refer	
treat	

9.2 Make word combinations using a word from each box. Look at C opposite to help you.

acutely
assessment
on
referral
waiting

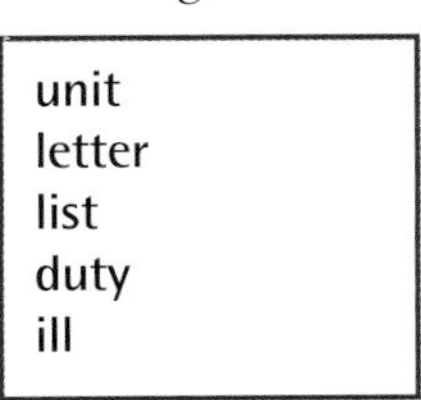
unit
letter
list
duty
ill

9.3 Which hospital departments would be most appropriate for the following patients? Look at A, B and C opposite to help you.

1 a woman in diabetic coma
2 a patient who has just had a radical prostatectomy
3 a patient who is to have a skin lesion removed
4 a man with a foreign body in his eye
5 a woman with a threatened abortion

9.4 Complete the extract from an information leaflet for patients. Look at A, B and C opposite to help you.

> **Information for outpatients**
>
> When you arrive at the (1) ...., please tell the receptionist who will welcome you, check your details, and direct you to the waiting area. The length of your visit will depend on the (2) .... you're going to have. You may need to have some (3) ...., such as an X-ray, which could mean going to another (4) .... Or you may be (5) .... to other professionals, such as a physiotherapist or dietician. You may need to revisit the clinic. If staff at the clinic want to see you again, another appointment will be arranged for you. If you need to be (6) .... to hospital for more treatment, either as an inpatient or for (7) .... surgery, you will be told when this is likely to happen.
> If you do not need further treatment you will be (8) .... to your GP's care.

Over to you

How do hospitals in your country organize admissions? How would you explain the procedure to a colleague from another country?

10 Primary care

A The National Health Service

The **National Health Service** (**NHS**) is responsible for health care for everyone in the UK, although a small number choose to pay for private care. Treatment is free but there is a **prescription charge** for drugs and **appliances**, such as a cervical collar, with exemptions for some patients, such as children and the elderly.

Primary care is provided by **general practitioners**, or **GPs**, (sometimes known as **family doctors**), nurses, dentists, pharmacists and opticians. GPs work in **practices** of 1 to 20. Practices are based in a **surgery** and look after the health of from 1,000 to 15,000 people in their local community. They also provide health education in areas such as smoking and diet, **run clinics**, give vaccinations, for example for influenza, and may **perform minor surgery** such as removal of warts and moles. If a patient needs specialist care, the GP will **make a referral** – **refer** the patient **to** a consultant in secondary care.

Patients are normally seen on an **appointment** basis. **Home visits** are **made** when patients are **housebound** – unable to leave their homes – or too ill to attend surgery. **Out-of-hours** (**OOH**) treatment, from 6 pm to 8 am, is provided by the local Primary Care Trust, which organizes shifts of GPs and **locum** GPs to cover if someone is absent.

Note: The noun **surgery** has three meanings:

- the building where GPs work – *The practice has moved to a new surgery on the High Street.*
- a time when GPs see patients – *Morning surgery is from 8.30 to 12.30.*
- the work of surgeons – *The patient needs urgent surgery on a burst appendix.*

B The practice team

A typical GP practice employs **receptionists.** They are responsible for initial patient contact, **making appointments**, taking requests for repeat prescriptions, **taking messages** from patients and other health care providers, and for **filing** and **scanning documents** into patient records. The **practice manager** has responsibility for finance and sometimes for IT, **supervises** reception **staff**, **hires locums**, and helps prepare the practice development plan. **Practice nurses** run asthma, diabetes, and cardiovascular disease clinics as well as **one-to-one** clinics for those who wish to give up smoking.

In addition to practice staff, GPs work with a number of health professionals (see Unit 8):

- **District nurses** visit temporarily housebound patients, such as recently **discharged** hospital patients, to **change dressings**, such as ulcer dressings.
- **Health visitors** visit families to **carry out check-ups** on young children – particularly under-fives – to make sure they're healthy. Special attention is paid to **families in need**, such as those living in poverty. They also do baby immunizations.
- **Midwives** run clinics for antenatal patients.
- **Physiotherapists** provide hands-on treatment but also teach patients exercises they can do to improve their condition after an accident or operation.

C A GP's day

Dr Stuart works in a practice in a small market town with three other family doctors. The surgery is in the centre of the town and is shared by three practices. This is a typical working morning when she is not the **duty doctor**, responsible for emergencies and urgent problems.

8.00 am	arrive at the surgery check the OOH email printout check for urgent and non-urgent messages
8.30 am	**check emails** from the health board and partners prepare for surgery
8.30 - 10.50 am	**morning surgery** (ten-minute appointments) 6 **pre-booked** last week 2 **booked** 48 hours **ahead** 4 **bookable on the day**
10.50 - 10.55 am	check with **Reception** for messages **sign prescriptions** and deal with **repeat prescription requests**
11.00 - 11.20 am	coffee break in the conference room with colleagues
11.20 - 11.30 am	check **home visit requests** and divide up visits with colleagues
11.30 am - 1.00 pm	home visits

10.1 Complete the sentences. Look at A opposite to help you.

1 Children, over-60s, and people with some chronic diseases do not have to pay .... in the UK.
2 Patients with mobility problems may be unable to go out. They are ....
3 The average GP .... is ten minutes long.
4 A .... is someone who takes the place of a staff member who is on leave.
5 Care outside working hours is known as ....................................-....................................-.................................... treatment.

10.2 Make word combinations using a word or phrase from each box. One word can be used twice. Look at A and B opposite to help you.

change
make
perform
refer
run
supervise
take

messages
staff
appointments
home visits
dressings
a clinic
a patient
minor surgery

10.3 Which member of a practice team would be responsible for each of the following? Look at B opposite to help you.

1 Running a clinic for pregnant women
2 Teaching a patient how to strengthen his broken leg
3 Letting the GP know that a patient can't come to her appointment
4 Running a clinic for people who want to lose weight
5 Visiting a patient who has just returned home after a hernia operation
6 Carrying out check-ups on children in a poor neighbourhood
7 Organizing cover for an absent doctor

10.4 Complete the diary for Dr Stuart's afternoon. Look at A and C opposite to help you.

1.00 – 2.00 pm	practice team meeting over sandwich lunch
2.00 – 4.00 pm	afternoon (1) .... 12 ten-minute (2) ....
4.00 – 4.20 pm	coffee break
4.20 – 5.00 pm	check with (3) .... for messages. Deal with home (4) .... and repeat (5) .... requests.
5.00 – 6.00 pm	paper work, e.g. (6) .... to secondary care, admin tasks, telephone calls to patients, private medical examinations
6.00 pm	phones switched to (7) .... service.

Describe a typical day for a GP in your country or in the country where you work.

11 Medical education 1

A Medical education in the UK

Medical education in the UK covers:

- **undergraduate** education – four or five years at **medical school**, the section of a university responsible for medical education
- a two-year **Foundation Programme** which provides training for new doctors after **graduation** through a series of placements in different specialties (see Unit 12)
- **postgraduate** training which doctors take to become GPs or **consultants** – senior specialists – often delivered through **colleges** for different specialties, for example the Royal College of Physicians
- **continuing professional development** in the form of courses and seminars, which doctors undertake throughout their working lives to keep up to date.

B Extract from an undergraduate prospectus

The MBChB (Bachelor of Medicine, Bachelor of Surgery) is a five-year undergraduate medical degree course. Most of your learning takes place in small groups. The main components are:

Core (Years 1–3)

An integrated programme of clinical and scientific topics mainly presented through **problem-based learning (PBL)**, where you work with others on a series of case problems.

Student Selected Modules

Student selected modules (SSMs) allow you to choose from a menu of subjects such as Sports Medicine or even study a language as preparation for an **overseas elective**, a hospital attachment of your own choice, between Years 4 and 5.

Vocational Studies and Clinical Skills

This component prepares you for the **clinical skills** required for contact with patients from Year 1 of your course, through periods of practical training where you are attached to a hospital department or general practice.

Clinical Attachments (Years 4 and 5)

A series of four-week **clinical attachments** in Medicine, Surgery, Psychological Medicine, Child Health, Obstetrics & Gynaecology and General Practice.

C A student's view

Ellen, a medical student, describes her course.

'I'm just finishing my first year of Medicine. What I like about this course is that you're involved with patients from the very beginning. Even in our first year, we spend time in hospital. Much of the course is PBL. We have two 2-hour sessions a week where we work in groups of eight to ten solving clinical problems. We decide together how to tackle the problem, look up books and online sources, make notes and discuss the case together. It's a great way of learning and getting to know the other students. In the past, medical students had **lectures** with the whole class taking notes from lecturers from 9.00 to 5.00, but now it's mainly group work, although we do have some lectures and **seminars**, where we work in small groups with a tutor. I like all of it, even the **dissection**. We get to cut up **cadavers** from the second month of the course.'

11.1 Match these activities to the stages of medical education in the UK given in A opposite.

1 dissecting cadavers
2 keeping a log of surgical procedures observed and performed
3 working for four months in accident and emergency to experience this specialty
4 taking a four-week attachment in Obstetrics and Gynaecology
5 taking an online course on recent developments in cardiovascular disease

11.2 Complete the sentences. Look at B and C opposite to help you.

1 Just before their final year, students have the chance to take an .... in a hospital of their choice anywhere in the world.
2 ....-.............................. .... contrasts with an approach where each subject is taught separately.
3 These days .... are often interactive, with regular opportunities for the students to ask questions.
4 In .... students learn how to treat and manage patients.
5 .... can be a topic from outside medicine, such as a foreign language.
6 Dissection of .... is an important part of the anatomy component.
7 The .... at the University of Edinburgh is one of the oldest in the UK.
8 She's a .... at the Royal; one of the leading paediatric heart specialists in the country.
9 The Royal .... of Surgeons in Edinburgh dates from 1505.
10 We have a .... each week where we discuss topics in a small group with our lecturer.

11.3 Match each of these activities to one of the components of the undergraduate course described in B and C opposite.

1 Julie spends six weeks working in a small hospital in the Himalayas.
2 A group of students discuss together the possible reasons for abdominal pain after meals in an obese 44-year-old male.
3 A small group of students trace the pulmonary artery in a cadaver.
4 Otto spends a month working in the paediatric ward of the local hospital.
5 Anne learns how to take blood from an elderly patient.
6 Juma chooses to study Travel Medicine in his fourth year.

Over to you

Describe the main components of your undergraduate course.

12 Medical education 2

A The Foundation Programme

The Foundation Programme is a two-year training programme which forms the bridge between university-level study at **medical school**, and specialist or general practice training. It consists of a series of **placements**, each lasting four months, which allow the junior doctor, known as a **trainee**, to sample different specialties, for example paediatrics. A year one trainee (**FY1**) corresponds to pre-registration house officer (**PRHO**) posts and a year two trainee (**FY2**) to senior house officer (**SHO**) posts. Each trainee has an **educational supervisor** who ensures that more senior doctors deliver training in different ways, including clinical and educational supervision. To progress, trainees have to **demonstrate** a range of **clinical competencies** which are **assessed** through observation in their workplace.

(*BMJ Careers* 2005; Amended with permission from the BMJ Publishing Group)

B People in medical education

tutor	An academic, or in some cases a postgraduate student, who leads tutorials.
demonstrator	In anatomy teaching, someone who **demonstrates how to** dissect. Demonstrators are often postgraduate students paying their way through medical school.
lecturer / senior lecturer	An academic with teaching and research responsibilities who contributes to the teaching of a particular discipline.
professor	A senior academic with teaching and research responsibilities for a particular discipline. Usually a leading figure in their discipline.
college tutor	A consultant responsible for delivering a college training programme.
clinical trainer	A consultant assigned to a trainee who **provides training** during periods of direct clinical care.
educational supervisor	A consultant who **supervises** a trainee's period of training.

C Medical qualifications

BMSc BMed Sci	**Bachelor of Medical Sciences.** A **degree** often **taken** after three years of medical studies by students who may wish to follow a career in medical research.
MBChB, MBBS BMBCh, BMBS	**Bachelor of Medicine, Bachelor of Surgery.** Bachelor degrees are undergraduate degrees. This is the first degree for UK doctors.
MD, DM	**Doctor of Medicine**
DRCOG	**Diploma of** the Royal College of Obstetrics and Gynaecology
MRCP MRCS	**Member of** the Royal College of Physicians or Member of the Royal College of Surgeons. Doctors become Members by successfully completing the assessment procedures in their **college.**
FRCS FRCS(Ed) FRCS(Glas) FRCSI	**Fellow of** the Royal College of Surgeons of England. Other colleges are indicated by the letters which follow, for example Edinburgh, Glasgow or Ireland. How doctors become a Fellow depends on their college. For the FRCS, further examinations must be passed. For other colleges it is by nomination or work assessment.

12.1

Complete the phrases with verbs from the box. Two phrases can be completed in two different ways. Look at A and B opposite to help you.

assess	deliver	demonstrate	provide	supervise	take

1 .... a competence or how to do something
2 .... a trainee by ensuring she successfully completes her training
3 .... a course or a training programme (as a teacher)
4 .... a course or a training programme (as a student)
5 .... progress or competence

12.2

Match the two parts of the sentences. Look at A, B and C opposite to help you.

1 An FY1 is a doctor
2 A demonstrator is an anatomy teacher
3 A clinical trainer is a consultant
4 A supervisor is a consultant
5 A medical school is
6 A placement is
7 A college is
8 A Fellow is a specialist

a a body of specialists responsible for delivering and assessing training in their specialty.
b responsible for the training programme of a trainee.
c a period spent as a trainee in a hospital or in General Practice.
d in the first year of the Foundation Programme.
e who has reached the highest level in their specialty.
f who provides training during periods of direct clinical care.
g part of a university responsible for medical education.
h who teaches dissection.

12.3

Write in full the qualifications of the doctors and surgeons. Look at C opposite to help you.

1 Mr A. H. Younghusband, MBChB, FRCS, FRCSI

2 Dr C Doyle, BMed Sci, DM, MRCP

3 Ms E Inglis, MBBS, FRCS

4 Dr E Merryweather, BM, MD, FRCP

Over to you

How do you become a specialist in your country? List the stages.

13 The overseas doctor

A Types of registration

To manage and treat patients in the UK, all doctors must **register with** the **General Medical Council** (**GMC**). There are several types of registration:

- **Provisional registration** is for doctors who have just qualified from medical school in the UK or from certain European Economic Area (EEA) member states.
- **Full registration** is for doctors who have completed their year's clinical training.
- **Limited registration** is for international medical graduates who have not completed the equivalent of a year's clinical training in the UK.
- **Specialist registration** is for doctors who have completed specialist medical training and have a Certificate of Completion of Training (**CCT**).

The GP Register is a register of all those eligible to work in general practice in the NHS.

Note: For full details of the General Medical Council see www.gmc-uk.org

B PLAB

Before they can **obtain full registration**, some categories of overseas doctors are required to take the **Professional and Linguistic Assessments Board** (**PLAB**) test. PLAB is designed to ensure those who pass can practise safely at the level of an SHO in a first appointment in a UK hospital.

Part 1 consists of a written test of knowledge, skills and attitudes. Part 2 is an **Objective Structured Clinical Examination** (**OSCE**). It consists of 16 five-minute clinical **scenarios**, known as **stations**, to assess professional skills.

Note: For a full description of PLAB, see http://www.gmc-uk.org/doctors/plab/

C PLAB stations and advice

OSCEs assess these skills:

- Clinical examination: Your ability to carry out a physical examination of a simulated patient, an actor trained to play this role, will be assessed. Uncomfortable or intimate examinations will be carried out using a **manikin**, an anatomical model.
- Practical skills: You will be assessed on practical skills such as suturing and giving intravenous injections.
- Communication skills: Your ability to interact with a simulated patient, or in some cases the examiner, will be assessed. Skills tested may include breaking bad news and giving advice on lifestyle.
- History taking: Your ability to take an accurate history and make a reasoned diagnosis will be assessed.

Advice on the stations from a successful candidate:

Read the instructions outside each station carefully. You have one minute for this.

Don't forget the **ABC** (airways, breathing, circulation) **protocol** in every emergency station.

Keep in mind **safety precautions** like throwing the **sharps** in the **sharps bin**.

Check the patient understands what is happening; then ask them about any concerns they may have. Don't just give a lecture. Listen carefully to what the actor says.

Note: Sharps are needles and blades which must be disposed of safely in a special container called a **sharps bin**.

13.1 What kind of registration might these doctors obtain? Look at A opposite to help you.

1 A newly qualified Spanish doctor

2 A newly qualified Nigerian doctor

3 A doctor who has successfully completed the first Foundation Year (FY1)

4 An SHO who has successfully completed the Foundation Programme and gained a Certificate of Completion of Training after several specialist registrar posts

13.2 Write the abbreviations in words. Look at A, B and C opposite to help you.

1 Any doctor who wants to work in the UK must register with the *GMC*.

2 Some overseas doctors must pass the *PLAB* test before they can register.

3 Part 2 of the test consists of an *OSCE*.

4 In any emergency, remember the *ABC* protocol.

5 Before you can obtain specialist registration, you must have a *CCT*.

13.3 Complete the text. Look at A, B and C opposite to help you.

My name's Musa and I come from Yemen. I came to the UK about two years ago, after graduating. Because Yemen is outside the EEA, I could only obtain (1) .... with the (2) .... at first. It was very difficult for me to obtain a place on a Foundation Programme. Although I speak good English, I had to take the (3) .... test to show that I could work safely in the UK. If I had to give some advice to other candidates, it would be that at counselling (4) .... , you shouldn't simply memorize a set of phrases. It's better to really think about what you're saying to the actor and get the intonation right.

After completing my year's clinical training, I was able to obtain (5) .... But I found it difficult to get an SHO post in my chosen specialty, paediatrics, as hospitals now have to demonstrate there isn't a suitable candidate from the EEA. Once I've completed my second Foundation Year, I should obtain a (6) .... which will allow me to proceed to (7) .... with the GMC, an important step on the road to becoming a paediatric consultant.

Over to you

Explain how a foreign doctor can register to work in your country. Find out how you can register to work in another country of your choice.

14 Symptoms and signs

A Describing problems

The problems which a patient reports to the doctor are called **symptoms**, for example pain or nausea. **Signs** are what the doctor finds, also known as **findings**, on examining the patient, for example high blood pressure or a rapid pulse rate. Symptoms are also known as **complaints**. To report a patient's symptoms or complaints, doctors say:

Mr Farnsworth was admitted **complaining of** chest pain.

In case notes, the abbreviation **c/o** is used:

c/o chest pain

B Presentation

Patients say they **went to (see) the doctor**; doctors say the patient **presented**. The symptom which causes a patient to visit a doctor – or to **present** – is called the **presenting symptom**, **presenting complaint** or **presentation**.

His	presenting symptom presenting complaint	was chest pain.

He **presented to** his GP **with** chest pain.

The usual presentation is chest pain.

C Talking about symptoms

Symptom	Meaning	Patients say
tiredness **lethargy** **fatigue** **lassitude**	loss of energy	I **feel tired** all the time. I **feel** completely **worn out**. Lately I've been feeling completely **exhausted** at the end of the day.
malaise	general feeling of being unwell	I feel unwell. I don't feel well. I've been **feeling off-colour** for two days. I **haven't been feeling myself** for a week. I've been **out of sorts** all day.
anorexia	loss of appetite	My appetite is very **poor**. I've been **off my food** for days.
weight gain	increase in weight	I've **put on** eight kilos in the last year. I've **gained** five kilos.
weight loss	decrease in weight	I'm not eating any less than usual but I've **lost** a lot of weight recently.
constipation	hard, infrequent faeces	My **motions** are very hard. I've been quite **constipated** lately. I'm not very **regular**.

Note: The verb **feel** is also used with other adjectives, such as **hot**, **cold**, **nervous**, **anxious**, **dizzy**, **weak** – *She said she felt dizzy.*

14.1 Complete the table with words from C opposite. Put a stress mark in front of the stressed syllable in each word. The first one has been done for you.

Noun	Adjective
ex'haustion	
fatigue	
lethargy	
tiredness	

14.2 Make word combinations using a word from each box. Look at A, B and C opposite to help you.

complain off- out of present put worn	with of out on colour sorts

14.3 Complete the sentences with the correct form of the verb *present*.

1 A 67-year-old man .... with a 9-month history of increasing shortness of breath.
2 The most common .... is loss of consciousness.
3 Cranial arteritis may .... as fever without any obvious causes.
4 The patient usually .... with a severe sore throat.
5 The .... symptoms in this patient could perhaps be due to renal failure.
6 Other conditions with a similar .... include acute cholecystitis.
7 Reduced growth is an important .... complaint of coeliac disease.
8 Two months following .... , the patient was able to walk.

14.4 Read the patient's description of her symptoms then complete the case report. Look at C opposite to help you.

I was well until a few months ago. In the beginning, I just felt off-colour and a bit tired. But lately I've been feeling completely worn out at the end of the day. I'm not eating any more than usual but I've put on nine kilos in the last year. My motions are hard and my hair has started to fall out.

Case 13

A 50-year-old housewife, who had been well until four months previously, (1) .... of tiredness and (2) .... She had (3) .... 9 kg in weight in the year before she (4) .... to her GP although she denied eating more than usual. She was (5) .... and she noticed that her hair had started to fall out.

Over to you

Write a short case report about this 60-year-old man:

I <u>haven't been myself</u> for several months now. I feel <u>completely worn out</u> after doing anything. I've <u>been off my food</u> and I've lost ten kilos in weight.

Write in the past tense and use medical terms for the underlined expressions. Practise writing similar case reports for your own patients.

15 Blood

A Full blood count

In the investigation of blood diseases, the simplest test is a **full blood count** (**FBC**). A full blood count measures the following in a sample of blood:

- the amount of haemoglobin
- the number of the different cells – **red blood cells** (erythrocytes), **white blood cells** (leucocytes) and **platelets** (thrombocytes)
- the volume of the cells
- the erythrocyte sedimentation rate (**ESR**) – a measurement of how quickly red blood cells fall to the bottom of a sample of blood.

B Anaemia

Anaemia is one of the commonest diseases of the blood. It may be due to:

- **bleeding** – loss of blood
- excessive destruction of red cells
- low production, for example because the diet is lacking, or **deficient in**, iron (Fe).

A medical student has examined an elderly patient with a very low level of haemoglobin and is discussing the case with her professor:

Professor: What's the **most likely diagnosis** in this case?
Student: Most probably carcinoma of the bowel with **chronic blood loss.**
Professor: What's **against** that as a diagnosis?
Student: Well, he hasn't had any change in his bowel habit, or lost weight.
Professor: What else would you **include** in the **differential diagnosis** of **severe anaemia** in a man of this age?
Student: He might have leukaemia of some sort, or **aplastic anaemia**, but that's **rare** – it would be very unusual. Another cause is **iron deficiency**, but he seems to have an **adequate** diet.
Professor: OK. Now, there's another cause of anaemia which I think is more likely.
Student: Chronic bleeding ulcer?
Professor: Yes, that's right. But what about **pernicious anaemia**? Can you **exclude** that?
Student: Well, he's got none of the typical neurological symptoms, like paraesthesiae.

C Pernicious anaemia

Jordi Pons, the medical student from Barcelona, has made some language notes in his textbook.

Pernicious anaemia (**PA**) is a condition in which there is atrophy of the gastric mucosa with consequent failure of intrinsic factor production and vitamin B_{12} malabsorption. The **onset** is **insidious**, with **progressively increasing** symptoms of anaemia. Patients are sometimes said to have a lemon-yellow colour owing to a combination of **pallor** and **mild jaundice** caused by excessive **breakdown** of haemoglobin because of ineffective red cell production in the **bone marrow**. A red **sore** tongue (glossitis) is sometimes present. Patients present with **symmetrical** paraesthesiae in the fingers and toes, early loss of **vibration sense**, and **progressive** weakness and ataxia. The spleen may be **palpable**.

onset = beginning
insidious = slowly developing
pallor = lack of colour
mild = slight
jaundice = bilirubinaemia
breakdown = division into smaller parts
bone marrow = soft tissue in the cavity of bones
symmetrical = each side the same
vibration sense = ability to feel vibrations
progressive = continuing to develop
palpable = can be felt with the hand

15.1 Find words in the box with opposite meanings. Look at B and C opposite to help you.

adequate	unlikely	mild	common	insidious	for
against	severe	rare	sudden	inadequate	likely

15.2 Make word combinations using a word from each box. Look at B and C opposite to help you.

bone	diagnosis
differential	sense
insidious	onset
iron	marrow
pernicious	increasing
progressively	deficiency
vibration	anaemia

15.3 Complete the sentences. Look at A, B and C opposite to help you.

1 A 39-year-old man presented with a history of .... abdominal distension over a period of six months.
2 Blindness may be caused by vitamin A ....
3 The bleeding and purpura are caused by abnormal .... function.
4 The white cell count is normal so we can .... acute leukaemia.
5 The yellow colour of her skin and conjunctivae is probably due to ....
6 There was a .... mass in the right upper quadrant of the abdomen.
7 Treatment is aimed at restoring fluid balance with .... intravenous fluids.
8 The anaemia may be due to increased red cell ....

15.4 Complete the conversation. Look at C opposite to help you.

Professor: What is against the diagnosis of pernicious anaemia on physical examination?
Student: The problem started quite suddenly. So it didn't have the typical (1) .... He doesn't have any skin (2) .... and he doesn't have (3) .... paraesthesiae, or absent (4) .... sense, and I couldn't feel his spleen.
Professor: What about his tongue?
Student: His tongue was normal and not inflamed or (5) ....

Over to you

List the causes of anaemia mentioned in the conversation in B opposite. Then choose another condition that you encounter regularly and make a similar list of the causes in English. Use the index to help you.

16 Bones

A Bones

Some common English names for bones:

English name	Anatomical name
skull	cranium
jaw bone	mandible
spine	vertebral column
breastbone	sternum
rib	costa
collarbone	clavicle
shoulder blade	scapula
thigh bone	femur
kneecap	patella
shinbone	tibia

B Fractures

A **fracture** is a break in a bone. Some of the different types of fracture:

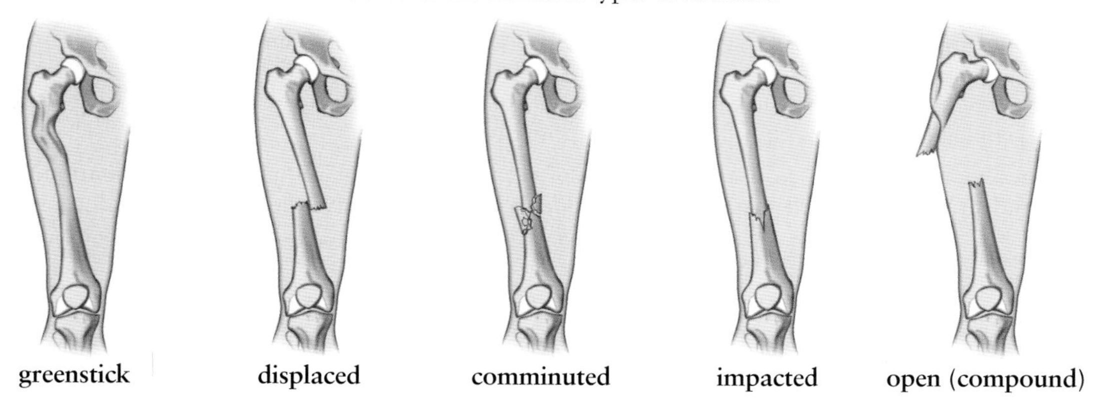

A **pathological fracture** is fracture in a diseased bone. A **fatigue** or **stress fracture** is due to repeated minor trauma, for example long-distance marching or running.

C Treatment of fractures

When the fragments of a broken bone heal and join together, they **unite. Union** may be **promoted,** or helped, by **reducing** the fracture – replacing the fragments in their anatomical position if they are displaced. After **reduction,** excessive movement of the broken bone is prevented by **fixation** – either external, for example a **splint** or **plaster of Paris cast,** or internal, for example a **pin** or a **plate and screws.** A displaced fracture which is not reduced may result in **malunion** – incomplete or incorrect union.

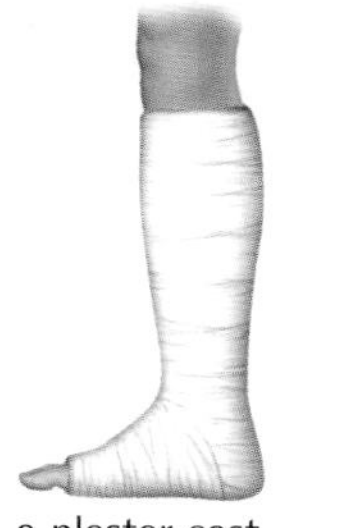
a plaster cast

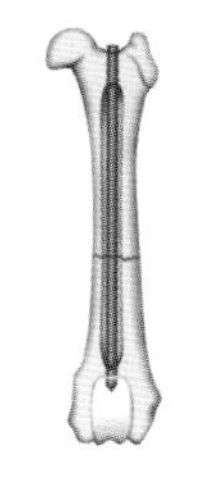
a pin

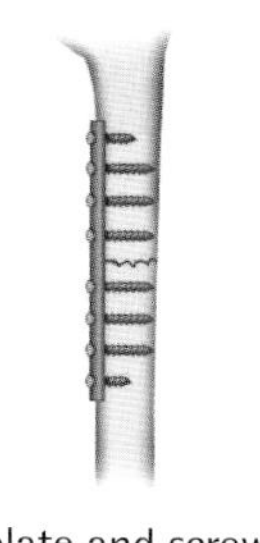
a plate and screws

Note: The verb **reduce** has several meanings in medicine:

- to make smaller – *I think we'd better reduce the dose of your tablets.*
- (in surgery) to return to anatomical position – *A hernia can normally be reduced by manipulation.*
- (in chemistry) to remove oxygen or add hydrogen – *Nitric acid is a reducing agent.*

16.1 Label the diagram using words from the box. Look at A opposite to help you.

breastbone
collarbone
jaw bone
kneecap
rib
shinbone
shoulder blade
skull
spine
thigh bone

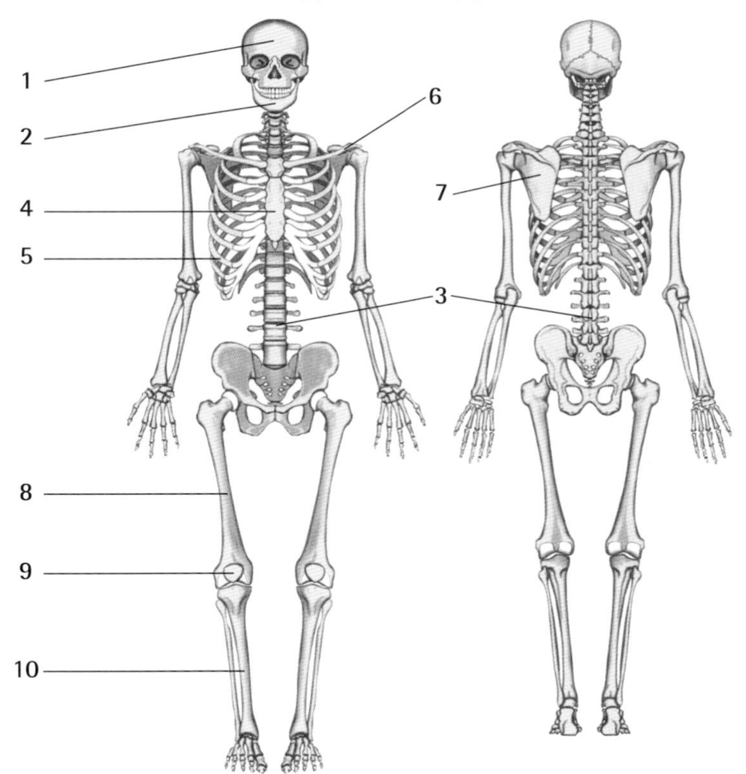

16.2 Match the types of fracture (1–5) with the descriptions (a–e). Look at B opposite to help you.

1 open
2 comminuted
3 displaced
4 greenstick
5 impacted

a There is a break in the skin.
b The bone is bent. It occurs mainly in children.
c The bone is broken into several pieces.
d The broken pieces are separated.
e The broken pieces are pushed together.

16.3 Complete the textbook extract. Look at C opposite to help you.

> (1) .... a fracture involves trying to return the bones to as near to their original position as possible. If a fracture is allowed to heal in a displaced position the fracture will (2) .... but it may go on to (3) ....

Over to you

You have diagnosed a stress fracture of the tibia in a young female dancer. How would you explain to her the cause and management of this condition?

17 Childhood

A Milestones

Childhood is the period during which a person is a child. It ends with **puberty** – the onset of sexual maturity. **Infant** is another word for a young child; **infancy** is the period from birth until about five years of age.

The **milestones** in a child's **development** and the ages at which they usually occur are:

- **sitting** – by 9 months
- **crawling** – by 12 months
- **first words** – by 18 months
- **walking** – by 18 months
- **talking** (two-word sentences) – by 3 years.

B Common infectious diseases

Disease	Common name
morbilli/rubeola	**measles**
rubella	**German measles**
varicella	**chickenpox**
infectious parotitis	**mumps**
pertussis	**whooping cough**
acute laryngotracheitis	**croup**
scarlatina	**scarlet fever**
rheumatic fever	**rheumatic fever**
tetanus	**lockjaw**
poliomyelitis	**polio**

C Coeliac disease

A medical student has made some language notes while reading her textbook.

Coeliac disease is a disease of the small intestine caused by **sensitivity to** gluten. It can present at any age but in infancy it appears after **weaning** on to cereals containing gluten. The **clinical features** include diarrhoea, **malabsorption** and **failure to thrive**. There may be signs of **malnutrition** and there may be some abdominal **distension**. There is **delayed growth** and **delayed puberty**, leading to short **stature** in adulthood.

sensitivity to = having a negative reaction to
weaning = changing the diet from milk only to solid foods
clinical features = the symptoms and signs of a disease
malabsorption = poor absorption
malnutrition = poor diet (nutrition) adjective = **malnourished**
thrive = grow strongly
distension = swelling
delayed = later than expected
failure = when something that is expected does not happen
stature = size, especially height

17.1 Complete the table with words from A and C opposite. Then complete the sentences with words from the table.

Verb	Noun(s)	Adjective(s)
delay		
develop		
distend		distended
fail		
nourish		

1 Babies with the fetal alcohol syndrome may present with .... to thrive.
2 Abdominal .... may be due to an enlarged liver.
3 Small amounts of alcohol in pregnancy can affect fetal ....
4 Mortality from measles can be reduced by better ....
5 .... in one or more of the milestones may be the first sign of disease.

17.2 Complete the sentences. Look at A and C opposite to help you.

1 After sitting, babies learn to .... and then to walk.
2 A child who has started eating solid food has been ....
3 Someone who is not very tall is said to be of short ....
4 The stages in a child's development are known as the ....
5 A child who is beginning to develop sexually has reached ....

17.3 Write the common English name for each disease, using your medical knowledge.

1 enlarged parotid glands
2 difficulty opening the mouth
3 rash and enlarged posterior occipital nodes
4 paroxysmal cough with vomiting
5 papules and vesicles, first on trunk
6 cough and cold followed by rash
7 sore throat and rash
8 swollen joints and a heart murmur
9 fever followed by muscle weakness
10 cough with stridor

Over to you

What are the main childhood illnesses in your country? What are the clinical features of those illnesses?

18 The endocrine system

A Excess and deficiency

An **excess** – too much, or a **deficiency** – too little, of circulating **hormones** causes a wide range of medical conditions, for example **hyperthyroidism** and **hypothyroidism**. Where there is an excess of hormone, one form of treatment consists of giving the patient something which **inhibits** the production of that hormone, as in the use of carbimazole to treat hyperthyroidism. When a hormone is **deficient**, treatment may be by **replacement therapy**, for example injections of insulin in the treatment of Type 1 diabetes.

Doctors say: Sufferers of type 1 diabetes are **deficient in** insulin.

B Negative feedback systems

1. TRH (thyrotrophin-releasing hormone) is **secreted** in the hypothalamus and **triggers** the **production** of TSH (thyroid-stimulating hormone) in the pituitary.
2. TSH **stimulates** the TSH receptor in the thyroid to increase **synthesis** of both T_4 (thyroxine) and T_3 (triiodothyronine) and also to **release** stored hormone, producing increased plasma levels of T_4 and T_3.
3. T_3 **feeds back** on the pituitary and perhaps the hypothalamus to inhibit TRH and TSH **secretion**.

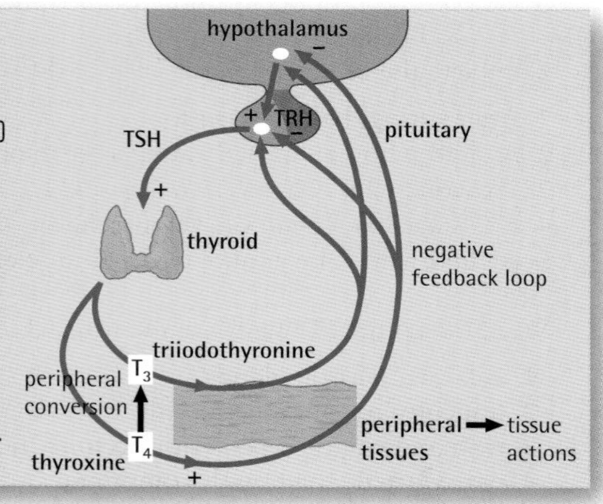

C Goitre

An enlarged thyroid gland is called a **goitre**. The enlargement may be **diffuse** – involving most of the gland, or **localized** – limited to a particular area, as in a **solitary** (single) nodule. The increased blood flow in diffuse enlargement, for example in Graves' disease, may give rise to a palpable **thrill** – vibration felt with the hand, and an audible **bruit** – noise heard through a stethoscope, over the gland.

D A letter of referral

Mrs Davis's doctor has referred her to an endocrinologist.

Dear Doctor,
I would be grateful if you would see this 50-year-old woman who has lost 20 kilos in weight in spite of eating more than usual. She describes herself as **overactive** and at first she thought the weight loss was due to this. But more recently she has developed **palpitations**, diarrhoea, and **heat intolerance**. She has noticed that her hands have a tendency to shake.
Her symptoms suggested hyperthyroidism and this was confirmed by my examination which revealed an enlarged thyroid, red sweaty palms and a **fine tremor** of the hands.

overactive: more active than is usual

palpitations: awareness of rapid or irregular heartbeat

heat intolerance: inability to cope with high temperatures

fine tremor: very slight involuntary movements

18.1 Complete the table with words from A, B and C opposite and related forms. Put a stress mark in front of the stressed syllable in each word. The first one has been done for you.

Verb	Noun
in'hibit	
produce	
release	
replace	
	secretion
	stimulation

18.2 Complete the passage from a textbook, using the illustration and your own knowledge. Look at B opposite to help you.

Pulses of GnRH (gonadotrophin-releasing hormone) are released from the hypothalamus and (1) .... LH and FSH (2) .... from the pituitary. LH (3) .... testosterone (4) .... from Leydig cells of the testis.
Testosterone (5) .... back on the hypothalamus/pituitary to (6) .... GnRH (7) .... FSH (8) .... the Sertoli cells in the seminiferous tubules to (9) .... mature sperm and the inhibins A and B. Inhibin causes feedback on the pituitary to decrease FSH (10) ....

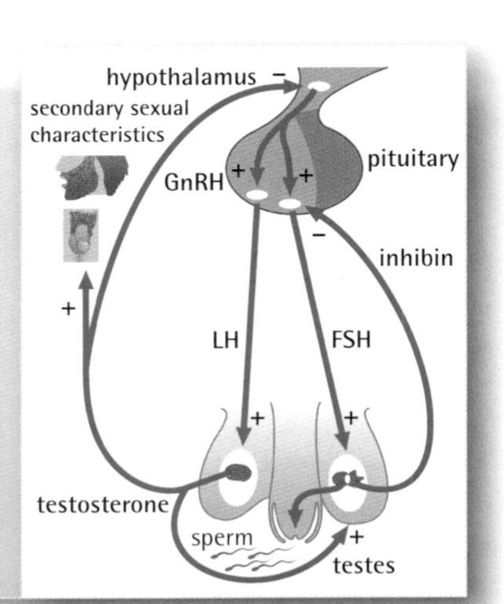

18.3 Complete the sentences. Look at A and C opposite to help you.

1 A .... change affects many parts of an organ or gland.
2 A .... change affects only one part.
3 His diet is .... in iron: he doesn't get enough iron.
4 .... T_3 and T_4 increase the basal metabolic rate.

18.4 Match Mrs Davis's symptoms (1–7) with the questions her doctor asked (a–g). Look at D opposite to help you.

1 diarrhoea
2 eating more
3 heat intolerance
4 overactivity
5 palpitations
6 weight loss
7 tremor

a Do you prefer hot weather or cold?
b Is your weight steady?
c What is your appetite like?
d Are your bowels normal?
e Are you able to sit and relax?
f Do your hands shake?
g Have you ever felt your heart beating rapidly or irregularly?

Over to you

Write a referral letter to an endocrinologist for a patient who you believe has hypothyroidism. Use the letter in D opposite as a model.

19 The eye

A Parts of the eye

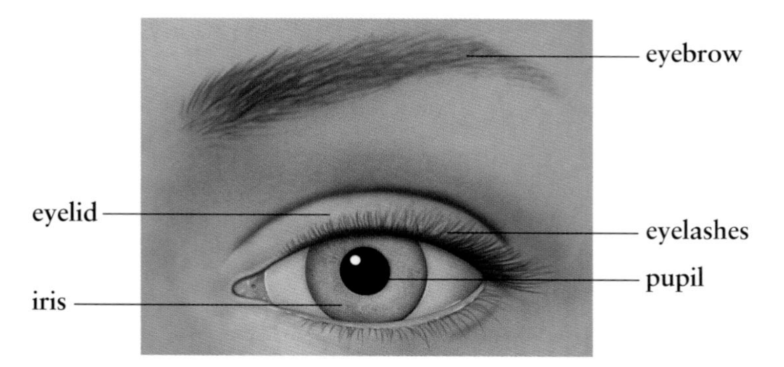

B Examination of the eye

Here is an extract from a textbook description of how to examine the eye.

Look for **squint** (strabismus), **drooping** of the upper lid (ptosis) or **oscillation** of the eyes (nystagmus). In **lid lag**, the upper eyelid moves irregularly instead of smoothly when the patient is asked to look down.

Next, examine the **pupils** and note whether:

- they are equal in size
- they are **regular in outline** (evenly circular)
- they are abnormally **dilated** (large) or **constricted** (small)
- they **react** normally to light and **accommodation** (focus on near objects).

To test the reaction to accommodation, ask the patient to look into the distance. Hold your finger in front of their nose, and ask the patient to look at it. The eyes should come together, or **converge**, and the pupils should **constrict** as the patient looks at the finger.

Check also for **cataract** (opacity of the lens).

C Retinopathy

Hypertensive changes in the retina can be classified from grades 1 to 4:

- grade 1 – **silver wiring** (increase in the light reflex) of the arteries only
- grade 2 – grade 1 plus arteriovenous **nipping** (indentation of veins where they are crossed by arteries)
- grade 3 – grade 2 plus **flame-shaped haemorrhages** and **cottonwool exudates**
- grade 4 – grade 3 plus papilloedema.

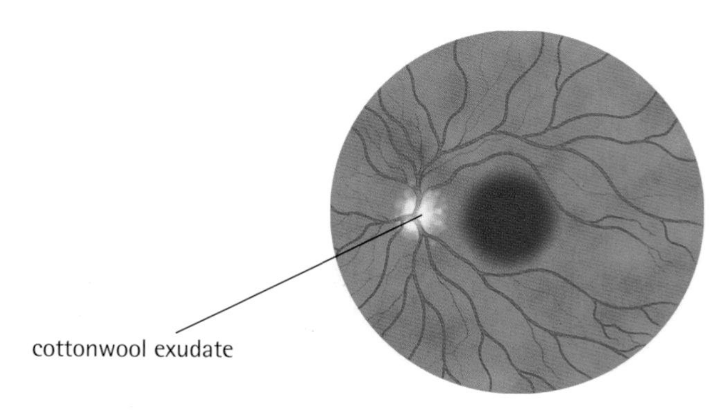

19.1 Complete the table with words from B opposite and related forms.

Verb	Noun	Adjective
accommodate		
	constriction	
	convergence	
	dilation, dilatation	
droop		
oscillate		
react		

19.2 Match the pictures (1–6) with the conditions (a–f). Look at B opposite to help you.

1
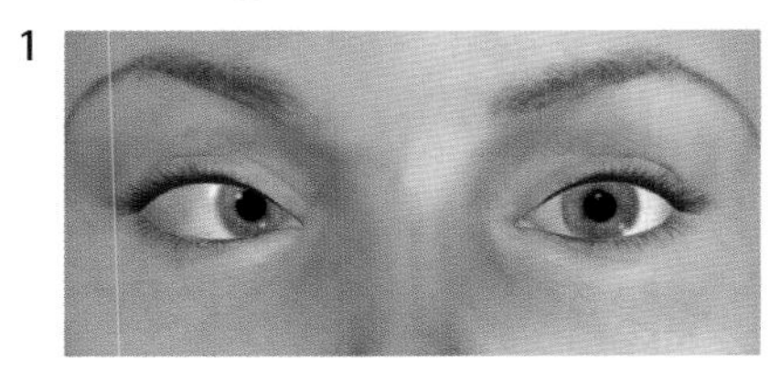

4
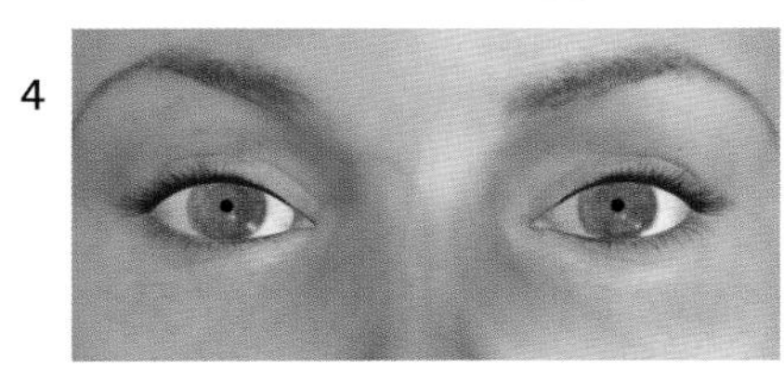

2
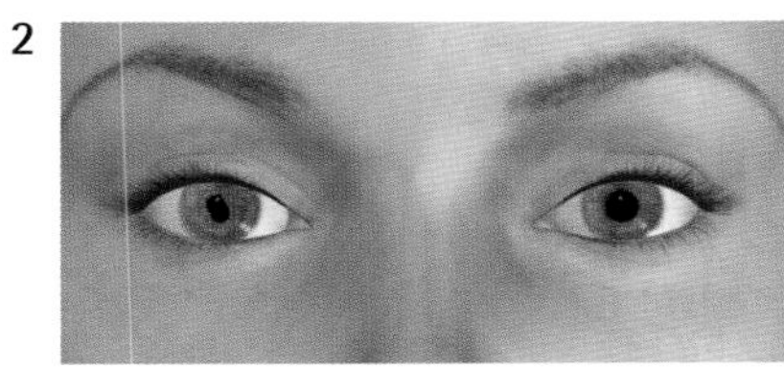

5
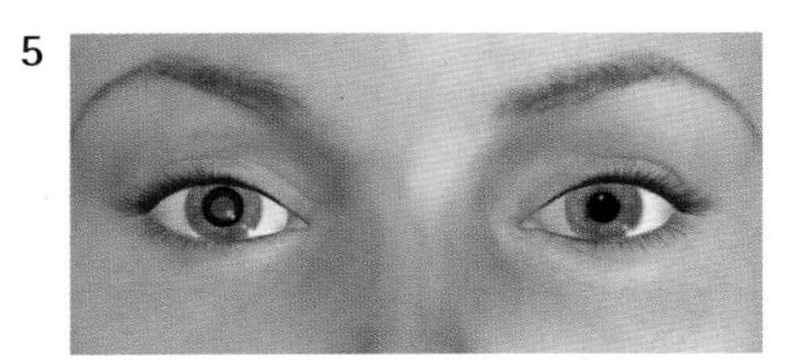

3
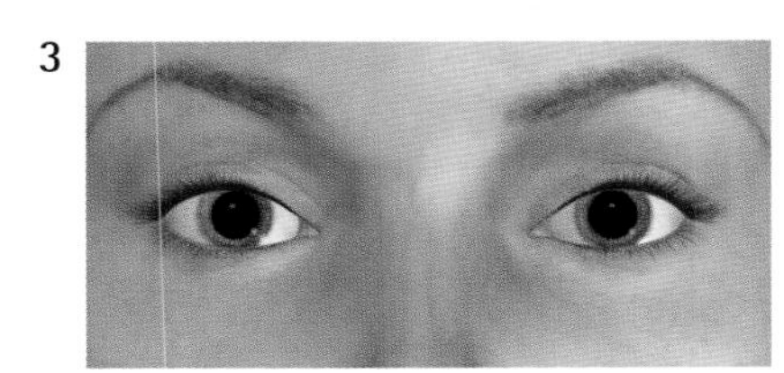

6
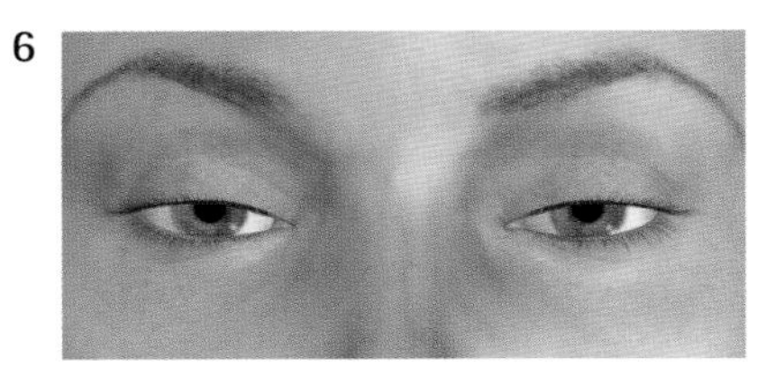

a drooping of lids
b dilated pupils
c irregular pupil
d cataract
e squint
f constricted pupils

19.3 Complete the extract from a textbook. Look at C opposite to help you.

Retinoscopy
Examine the retina with an ophthalmoscope, if possible with the (1) .... dilated to obtain the maximum view. Look for papilloedema, and for (2) ....– .... haemorrhages and (3) .... exudates. Assess the state of the (4) .... and note the presence of any narrowing, as well as (5) .... at arteriovenous crossings.

Over to you

Traditionally, eyesight problems are corrected with spectacles or contact lenses. In recent years, laser therapy has become a popular alternative. What are the advantages and disadvantages of this technique?

20 The gastrointestinal system

A Examination of the abdomen

Here is an extract from a textbook description of how to examine the abdomen.

> Note if the abdomen is **distended** by fluid or gas. The presence of fluid can be confirmed by demonstrating **shifting dullness**: **percuss**, or tap, first with the patient lying supine – flat on their back; then ask the patient to lie on one side and percuss again. If fluid is present, the dull note heard on percussion moves. **Palpate** each region, feeling for **tenderness** – pain when touched, or **masses** – palpable enlargement of tissue. Note also any **guarding** or **rigidity**, shown by contraction of the abdominal muscles. Guarding may be due to tenderness or anxiety and can be reduced if the patient is persuaded to relax. Rigidity, however, is constant and is due to peritoneal irritation. **Rebound tenderness** is pain when the palpating hand is suddenly removed. It is a sign of peritonitis. Listen for **bowel sounds**.

B The faeces

There are several words for the **faeces.**

Doctors sometimes say:

> There was blood in the **stools.**

> Have you **passed** black **stools**?

Patients sometimes say:

> My **motions** have been very loose lately.

Bowel movement is used to refer to defecation:

> Have your bowels moved today?

> Have you had a bowel movement today?

Bowel habit is a medical expression meaning the pattern of defecation.

> Have you noticed any **change of bowel habit**?

> How often do you **open your bowels**?

> Are you **going to the toilet** more often than normal?

Change in bowel habit could be **constipation** – hard, infrequent stools, or **diarrhoea** – frequent soft or liquid stools.

Normal stools are brown in colour, and semi-solid, or **formed.** The **consistency**, or degree of hardness and softness, can be shown on a scale:

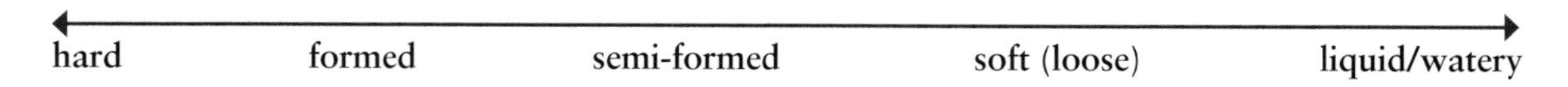

hard — formed — semi-formed — soft (loose) — liquid/watery

The colour can vary from black, due to altered blood as in melaena, to yellow, grey or even white. Melaena stools are often described as **tarry** – like tar, the black sticky substance that is used in road making. The stools may be red when fresh blood is present. Blood that can only be detected with special tests is called **faecal occult blood** (**FOB**). When there is a high fat content, the stools are pale, and are sometimes described as **clay-coloured.** Stools that are large in volume are described as **bulky**. A bad smell is described as **foul** or **offensive.**

20.1 Complete the case report. Look at A opposite to help you.

Case 14

Physical examination revealed a thin girl with slight pallor. She was not obviously dehydrated. The temperature was 38°C, pulse 100/min, blood pressure 110/80 mmHg. Examination of the rest of the cardiovascular and respiratory systems was normal. The abdomen was not (1) .... There was generalized (2) .... , which was most marked in the right lower (3) .... and was associated with (4) .... but not (5) .... There was no rebound (6) .... and no (7) .... were felt. (8) .... sounds were reduced.

20.2 Match the descriptions of the stools (1–6) with the conditions most likely to cause them (a–f), using your medical knowledge. Look at B opposite to help you.

Type of stools	Condition
1 loose, bloody	a gastric ulcer
2 loose, pale, bulky	b irritable bowel syndrome
3 clay-coloured	c ulcerative colitis
4 black, tarry	d cholera
5 small, hard	e coeliac disease
6 clear, watery with mucus	f obstructive jaundice

20.3 Match the features (1–7) to the doctor's questions (a–g). Look at B opposite to help you.

1 blood
2 bowel habit
3 change in bowel habit
4 bulk
5 colour
6 consistency
7 offensiveness

a How often do you open your bowels?
b Are you going to the toilet more often than normal?
c Are the motions hard or loose?
d Do the motions have an unusual smell?
e What about the appearance of the stools?
f Have you passed black stools?
g Is the size or the amount of the stool normal?

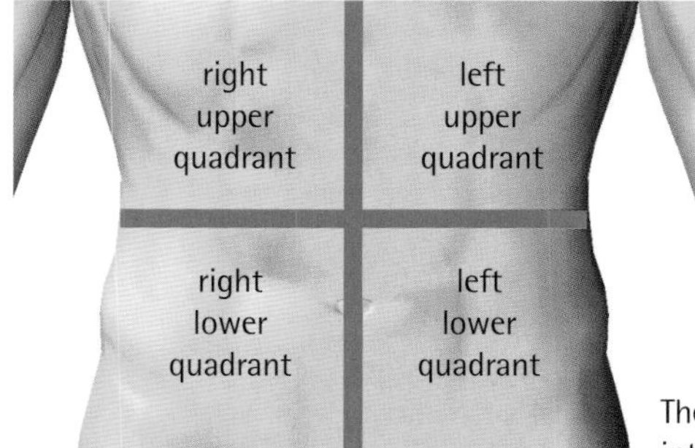

The abdomen can be divided into four quadrants.

Over to you

Look back at 20.2 above. In what other conditions that you encounter regularly is the appearance of the stools typical? How would you describe their appearance?

21 Gynaecology

A The female reproductive system

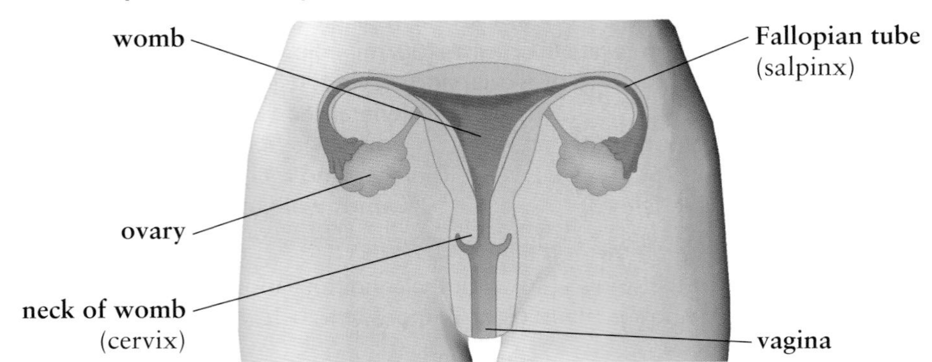

B Menstruation

A **period** is the common name for a **menstrual (monthly) period**. The **onset of menstruation** is known as menarche. The last menstrual period is commonly abbreviated in doctors' notes: **LMP 2/52 ago** means the last menstrual period was two weeks ago. The **menstrual cycle**, or length and frequency of periods, is usually written in the form 4/28, which means lasting 4 days and occurring every 28 days. If a period lasts more than four or five days it can be described as **prolonged.** The term **heavy periods** means excessive blood loss – menorrhagia, often with the **passage of clots** – coagulated blood. The term **period pains** means dysmenorrhoea, or painful menstruation.

The time when a woman stops menstruating, normally at about the age of 50, is called the **menopause** or **climacteric.** In everyday English it is known as the change of life, or simply **the change.** Symptoms of the menopause include **hot flushes** – sudden sensation of heat – and **night sweats.**

C A gynaecological consultation

A gynaecologist is talking to a 30-year-old woman.

Gynaecologist	Patient
Are your periods **regular**?	Yes.
How often do you **get** them?	Every four weeks.
How old were you when you started to get them?	About 12.
When was your **last period**?	A week ago.
How long do the periods last usually?	4 or 5 days.
Would you say they are **light** or **heavy**?	Light.
Do you **get clots**?	No.
Do you get **period pains**?	Not really.
Is there any **discharge** between the periods?	A little.
What colour is it?	White.

D Contraception

For women, methods to prevent pregnancy include the **oral contraceptive pill** (known as **the Pill**), the **diaphragm**, and the **intrauterine device** (IUD) or **copper coil. Condoms** are available for both men and women.

21.1 Write a simple English phrase for each of the medical terms below using your medical knowledge. Look at A and B opposite to help you.

1 hysterectomy
2 menorrhagia
3 salpingitis
4 cervical biopsy

21.2 Read the conversation between the gynaecologist and the patient in C opposite, and complete the notes about the patient.

menarche:
menstrual cycle:
LMP:
menorrhagia?
dysmenorrhoea?
discharge?

21.3 Now write the questions that the doctor asked. Look at C opposite to help you.

menarche:
menstrual cycle:
LMP:
menorrhagia?
dysmenorrhoea?
discharge?

21.4 Complete the case report. One word is needed twice. Look at B and C opposite to help you.

Case 15

A 45-year-old woman had been having (1) .... periods lasting for 8 days, with the passage of (2) .... , for 9 months. There was no bleeding between (3) .... or after intercourse. Her (4) .... were not particularly painful. She had not noticed any hot (5) .... or night sweats, and her general health had always been good. She had taken the (6) .... contraceptive (7) .... until a year previously, when a copper (8) .... was fitted. She had had a normal pregnancy when she was 25.

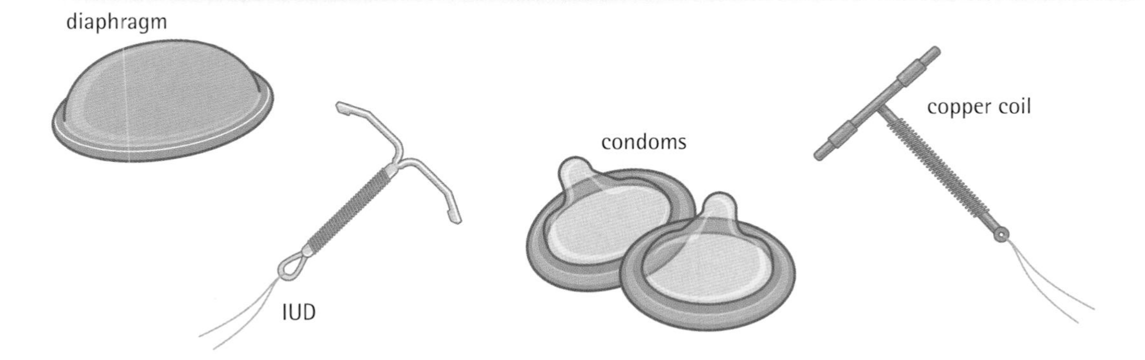

Over to you

What is the attitude to contraception in your country? At what age do you think females should be prescribed contraceptives?

22 The heart and circulation 1

A Shortness of breath

Shortness of breath, or **breathlessness**, is dyspnoea. At first this is caused by **exertion** – physical activity such as climbing stairs – but in severe cases it may be present even **at rest**. A patient who is breathless when lying flat (orthopnoea), for example in bed, will tend to sleep raised up on two or more pillows. The abbreviation SOBOE stands for **shortness of breath on exercise** (or **on exertion**, or **on effort**).

Patients say:

> I get terribly short of breath climbing stairs.

Doctors can ask:

> How many pillows do you sleep on?

B Heart rhythm

The normal **resting heart rate** is 65–75 **beats per minute**. In athletes it may be as low as 40 beats per minute. In extreme athletic activity, the heart rate can go as high as 200/min. The heart **rhythm** may be **regular** or **irregular**. In an irregular rhythm (arrhythmia), there may be early beats which interrupt the regular rhythm (**premature beats**); or the rhythm may vary with respiration; or it may be completely irregular, as in **fibrillation**. When patients are aware of irregularity, they describe the symptom as **palpitations**.

Case 4

A 22-year-old student was admitted to hospital with a long history of heart problems. She had been increasingly tired, with shortness of breath on exertion, orthopnoea, and palpitations. A **mitral valve replacement** had been carried out 3 years previously and this had stabilized the symptoms of heart failure but was followed by **episodes** (attacks) of **atrial fibrillation**, which had been particularly severe for the 6 months before admission.

C Heart failure

Heart failure occurs when the heart is unable to maintain sufficient **cardiac output** – the amount of blood pumped by the heart each minute – for the body's needs. It may involve the left side of the heart, the right side, or both. In **left heart failure** the main symptom is breathlessness. The symptoms of **right heart failure** include **peripheral oedema** (**swelling**), beginning in the feet and ankles. This is known as **pitting oedema** if, when a finger is pushed into the swelling, it causes a small depression or pit.

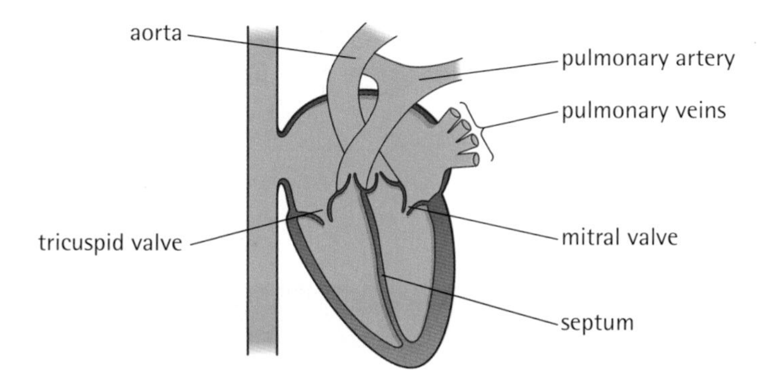

22.1 Complete the conversation based on the case history in B opposite.

Doctor: What seems to be the problem?
Patient: I've been getting (1) ....
Doctor: How long have you had them?
Patient: For about six months. But I've had heart problems for years, with tiredness and (2) .... of (3) .... In the end I couldn't walk more than a hundred metres without having to stop. I had to sleep on three (4) .... I had a (5) .... replacement three years ago, and that improved things for a while.

22.2 Make word combinations using a word from each box. Two words can be used twice. Look at B and C opposite to help you.

at
atrial
cardiac
heart
on
pitting
premature

output
failure
oedema
fibrillation
beats
effort
rest

22.3 Write the words a patient would use to describe the symptoms below. Look at A, B and C opposite to help you.

1 dyspnoea
2 arrhythmia
3 orthopnoea
4 oedema

22.4 Complete the case report. Look at A, B and C opposite to help you.

Case 13

A 60-year-old woman attended her GP's surgery complaining of breathlessness on (1) .... This had been increasing over the previous eight months until it was producing problems at around 500 metres walking on the level. There was no history of chest pain. She had had several (2) .... of fast (3) .... which lasted 20–30 minutes and were associated with some (4) .... of breath. She had noticed some (5) .... of her ankles by the end of the day. This disappeared overnight.

Over to you

How would you manage the treatment of the woman in 22.4 above?

23 The heart and circulation 2

A Physical examination

Medical examination is normally carried out in four stages: **inspection** (looking), **palpation** (feeling with the hands), **percussion** (tapping with a finger) and **auscultation** (listening with a stethoscope).

Note: The verb is **palpate**; the noun is **palpation** (not **palpitation** – see Unit 22).

B Examining the heart and circulation

Here is an extract from a textbook description of how to examine the cardiovascular system.

Look at the lips, tongue and nails for the blue discoloration of **cyanosis**. Cyanosis may be **central** or **peripheral**. **Inspect** the hands for **clubbing**. Feel the **radial pulse** at the wrist and note the **rate** (for example 70/min) and **rhythm** (**regular** or **irregular**). The pulse may be **irregular in force** as well as **time**. Check that the other **peripheral pulses** are **present**. Measure the blood pressure, and assess the **jugular venous pressure (JVP)**. **Palpate** the chest for the **apex beat** – the normal position is the fifth left **intercostal space**, one centimetre medial to the **midclavicular line**. Feel for any **thrills**. Heart size may be measured by percussion. Listen for **murmurs** and other abnormal sounds, for example **friction rubs**, beginning at the **mitral area**. Murmurs may be **soft** or **loud**. A **harsh** murmur is loud and rough.

Note the time of any murmur in relation to the cardiac cycle. The most common murmurs are:

- mid-systolic (in the middle of systole)
- pan-systolic (lasting for the whole of systole)
- early diastolic
- mid-diastolic
- late diastolic (pre-systolic)

Continue by listening at the tricuspid, aortic and pulmonary areas.

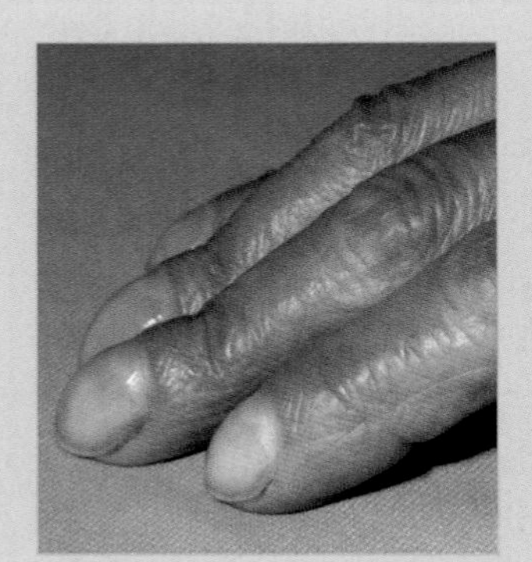

Finger clubbing

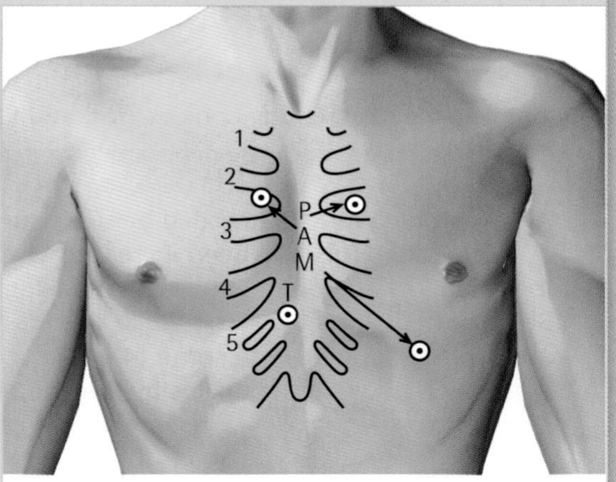

Areas of auscultation. The letters indicate the approximate position of the **heart valves**: P **Pulmonary valve**; A **Aortic valve**; M **Mitral valve**; T **Triscupid valve**. The circles indicate the position for auscultation for cardiac murmurs indicating **valvular heart disease**. The ribs are numbered.

23.1 Complete the table with words from A opposite. Put a stress mark in front of the stressed syllable in each word. The first one has been done for you.

Verb	Noun
'auscultate	
examine	
inspect	
palpate	
percuss	

23.2 Put the steps for examining the heart and circulation in order, according to the four-stage system. Look at A opposite to help you.

a Measure the heart size.
b Are there any murmurs?
c Feel the radial pulse.
d Look for clubbing.
e Locate the apex beat.
f Note any thrills.

23.3 A doctor is presenting the case of a 43-year-old woman at a meeting in the Cardiology Department. Complete the text of her presentation. Look at B opposite to help you.

> On examination she was pyrexial with a temperature of 38.5. She was short of breath. Her pulse was variable between 100 and 180 and was irregular in time and (1) ....
> Her blood pressure was 130/80 and her JVP was up 5 centimetres showing normal movement with respiration. Her peripheral (2) .... were all present and there was no (3) .... or (4) .... cyanosis. Her apex (5) .... was displaced to the anterior axillary line but still in the fifth intercostal (6) ....
> Her heart sounds were very interesting. When she was initially examined it was noted that she had pan-systolic and mid-diastolic (7) .... , heard best at the apex. When she was examined some hours later, there was a harsh pericardial friction (8) .... all over the precordium. Our diagnosis at that time was of mitral stenosis and incompetence with a recent onset of pericarditis and atrial fibrillation.

Over to you

How many signs of heart disease can you find in B opposite? Can you add any signs of heart disease to the list?

24 Infections

A Fever

A medical student has made some language notes on a case report.

Case 45

A 24-year-old man presented with a **fever** which he had had for three days. On the third day he had had a severe attack of fever with sweating and **rigors**. The only past history of relevance was hepatitis four years earlier and **glandular fever** (infection with Epstein–Barr virus) at the age of 18 years. He had returned from Africa three weeks previously.

fever = **pyrexia** (also remember **PUO – pyrexia of unknown origin**)
fever also known as **temperature** – 'I've got a temperature'.
adjectives = **feverish/febrile** and **pyrexial**
opposites = **afebrile/apyrexial**

Some symptoms of fever
sweating
rigors (severe shivering and sensation of coldness, also known as **chills**)

B Microorganisms

Infections differ from other diseases in a number of aspects:

- Most importantly, they are caused by living **microorganisms** – such as viruses or bacteria – that can usually be identified, thus establishing the aetiology early in the illness. Many of these organisms, including all bacteria, are sensitive to antibiotics and most infections are potentially **curable**, unlike many **non-infectious** degenerative and chronic diseases.
- **Communicability** is another factor which differentiates infections from non-infectious diseases. **Transmission** of pathogenic organisms to other people, directly or indirectly, may lead to an **outbreak** or epidemic.
- Finally, many infections are preventable by hygienic measures, by vaccines, (especially live attenuated vaccines such as rubella vaccine) or by **drug prophylaxis** (for example, chloroquine to prevent malaria).

Microorganisms include bacteria, viruses, fungi, protozoa (such as the **parasite** that causes malaria). Another general word for these pathogens is **microbes**. Patients often refer to microbes as **germs** or **bugs**.

Notice the common expressions for acquiring an **infectious disease**:

Could he have	caught picked up	some disease from the dog?

I think I've caught the flu bug that's going round.

C Source and spread of infection

Here is an extract from a medical textbook.

Infection may originate from the patient (**endogenous**), usually from skin, nasopharynx or bowel, or from outside sources (**exogenous**), often another person who may be either suffering from an infection or **carrying** a pathogenic microorganism. **Carriers** are usually healthy and may harbour the organism in the throat (for example, diphtheria), bowel (salmonella), or blood (hepatitis B or HIV). Non-human sources of infection include water (e.g. cholera), milk (e.g. tuberculosis), food (e.g. botulism), animals (e.g. rabies), birds (e.g. psittacosis) and also the soil (e.g. legionella – **Legionnaires' disease**).

The **incubation period** is the period between the invasion of the tissues by pathogens and the appearance of clinical features of infection. The **period of infectivity** is the time that the patient is infectious to others.

24.1 Match the two parts of the sentences. Look at A, B and C opposite to help you.

1 1988 saw the UK launch of live attenuated
2 Chickenpox (varicella) is a common infectious
3 Rabies has an incubation
4 The patient remained febrile
5 He was admitted with a four-day history of influenza-type symptoms of fever with
6 Quite a proportion of patients who recover from hepatitis B
7 The central part of Africa is in the midst of an epidemic
8 Measles (rubeola) is most
9 Lyme disease is caused by transmission
10 PUO stands for

a period ranging from four days to many months.
b rigors, myalgia and general malaise.
c become carriers of the virus.
d infectious during the catarrhal stage.
e disease of childhood.
f of AIDS.
g of B. burgdorferi from animal to man by ixodid ticks.
h with peaks of temperature of 39.5°C.
i pyrexia of unknown origin.
j measles, mumps, and rubella (MMR) vaccine.

24.2 Complete the case report on the patient in A opposite. Look at A, B and C opposite to help you.

Case 45

On examination, he looked unwell. His pulse rate was 100/minute. He had a palpable spleen. The combination of (1) .... and (2) .... in a patient who has recently returned from Africa strongly suggests a diagnosis of malaria. The (3) .... period is usually 10–14 days. In this case, the patient admitted he had not been taking (4) .... regularly. The diagnosis was confirmed by the presence of (5) .... in his blood film.

24.3 Complete the sentences. Look at A, B and C opposite to help you.

1 An infection which can be treated successfully with antibiotics is ....
2 Another word for an epidemic is an ....
3 Bacteria and viruses are examples of ....
4 Someone whose temperature is normal is ....
5 The common infection with Epstein–Barr virus is known as ....

Over to you

Cases of HIV infection reach record high in the UK

The Times, 25 November 2005

Describe the situation with regard to HIV in your country. What measures are being taken to control it?

25 Mental illness

A Psychiatric disorders

Psychiatric disorders can be divided into **organic** and **functional. Dementia** is a mental disorder due to organic brain disease. The commonest form of dementia is that associated with old age: **senile dementia.** Disorders in which there is no obvious pathology or anatomical change in an organ are termed **functional.** These are described below.

B Substance abuse

Abuse of a substance means using it in a way that is harmful. The commonest forms of substance abuse are **alcoholism** and **drug abuse.**

C Affective disorders

Here is an extract from a medical textbook.

Affect and **mood** are similar in meaning and refer to the emotions (for example, happiness or sadness). Affect tends to be used for temporary emotions, and is expressed through manner of speaking, facial expression, or behaviour. Mood is used to refer to a more permanent emotional state. The most common form of affective disorder is **depression**, the symptoms of which are:

- poor appetite or significant weight loss
- **sleep disturbance** (for example, insomnia – inability to sleep)
- **fatigue** (loss of energy)
- **psychomotor agitation** (excessive movement and thought) or **psychomotor retardation** (slowing of movement and thought)
- **loss of interest** in stimulating activities
- decreased ability to think and concentrate
- feeling that one is of no value to others, or that one has done something wrong
- recurrent thoughts of death or **suicide.**

Five, or possibly four, of the above symptoms, occurring nearly every day for at least two weeks, constitute a **major depression.**

D Neurotic and stress-related disorders

An example of neurotic disorder is **obsessive compulsive disorder.** An **obsession** is an idea that is so persistent that it interferes with the patient's life. A **compulsion** is an obsessive idea that forces the patient to act even though they recognize that it is unnecessary. A common form of this is compulsive washing of the hands. **Stress** is a feeling of being unable to cope. It can lead to **anxiety** or fear of problems. A sudden attack of anxiety is called a **panic attack.**

E Other types of functional disorder

These include:

- **behavioural syndromes** associated with physiological disturbance (such as **eating disorders**)
- disorders of adult personality and **behaviour** (for example, **personality disorder**)
- **mental retardation** – delayed mental development
- schizophrenia and other disorders in which there are **delusions** (false beliefs).

25.1 Complete the table with words from A, C, D and E opposite.

Noun	Adjective
	affective
	anxious
	behavioural
	demented
	disturbed
	suicidal

25.2 Make word combinations using a word from each box. Two words can be used twice. Look at B, C and E opposite to

behavioural eating major mental personality psychomotor sleep substance	abuse disturbance retardation disorder depression syndrome

25.3 Complete the sentences. Look at A, C, D and E opposite to help you.

1 The way a person behaves is his or her ....
2 A persistent emotional state is a ....
3 A sudden attack of anxiety is a ....
4 The form of dementia associated with ageing is called ....
5 A disorder which is not associated with pathological changes is ....
6 An idea which forces a patient to repeat unnecessary actions is a ....

25.4 Which symptoms of depression was this patient suffering from? Look at C opposite to help you.

Case 41

A 56-year-old woman presented to her GP complaining of increasing tiredness over the past few months. She had lost interest in most things. She was sleeping poorly and tended to wake up early, but denied any suicidal tendencies. She was thirsty and was passing urine more often. She was eating normally and her weight was steady.

Over to you

Do you think the woman in 25.4 above was suffering from major depressive illness? Give your reasons.

26 The nervous system 1

A Sensory loss

The central nervous system controls the **sensory** and **motor functions** of the body. Diseases of this system therefore lead to loss of some of these functions.

Function	Loss	Other symptoms
hearing	**deafness**	**buzzing** or **ringing** in the ear (tinnitus)
sight	**blindness**	**double vision** (diplopia) **blurring** (loss of **visual acuity** – clarity of vision)
sensation (feeling)	**numbness** (anaesthesia)	**tingling** or **pins and needles** (paraesthesiae)
balance	**unsteadiness** (ataxia)	**dizziness** (vertigo)

Note: There are no common words for loss of, or conditions relating to, taste and smell.

B Motor loss

Motor loss symptoms and signs include:

- **weakness** – loss of power
- **paralysis** – complete loss of power
- **tremor** – involuntary rhythmic movement, especially of the hands
- **abnormal gait** – unusual manner of walking.

Speech may also be affected, for example with **hoarseness** – a rough, deep voice as in vocal cord paralysis. **Slurred speech** means poor articulation, as in cerebellar disease.

C Loss of consciousness

Patients may describe sudden loss of consciousness in a number of ways:

I	**passed out.** had a **blackout.** **fainted.**

I had a	fit. seizure. convulsion.

Fit, seizure and **convulsion** are all used to refer to violent **involuntary movements**, as in epilepsy.

Doctors may say: When did you lose consciousness?

Here is a passage from a textbook on the causes of loss of consciousness.

> The principal differential diagnosis is between an **epileptic fit** and a **syncopal attack**, or **fainting**. **Syncope** is a sudden loss of consciousness due to temporary failure of the cerebral circulation. Syncope is distinguished from a seizure principally by the circumstances in which the event occurs. For example, syncope usually occurs whilst standing, under situations of severe **stress**, or in association with an arrhythmia. Sometimes a convulsion and **urinary incontinence** – **loss of control** of the bladder – occur even in a syncopal attack. Thus, neither of these is specific for an epileptic attack. The key is to establish the presence or absence of **prodromal symptoms**, or symptoms that occur immediately before the attack. Syncopal episodes are usually preceded by symptoms of **dizziness** and **light-headedness**. In epilepsy, people may **get a warning**, known as an **aura**, that an attack is going to happen.

Note: The noun **convulsion** is often used in plural form – *He had* ***convulsions*** *as a child.*

26.1 Complete the table with words from A, B and C opposite.

Adjective	Noun
blind	
conscious	
deaf	
dizzy	
numb	
light-headed	
unsteady	

26.2 Make word combinations using a word from each box. Look at A, B and C opposite to help you.

double epileptic prodromal syncopal urinary visual	acuity attack incontinence symptom vision fit

26.3 A doctor is trying to determine the cause of loss of consciousness in a 52-year-old man. Complete the doctor's questions. Look at C opposite and at the table in 26.1 above to help you.

Did you lose (1) .... suddenly or gradually?

Did you get a (2) .... of the attack?

What were you doing before you (3) .... out?

Were you worried or under any (4) .... at the time?

Did you feel (5) .... or (6) ....-.................................... before the attack?

Did you lose (7) .... of your bladder?

Did your wife notice any (8) .... movements while you were unconscious?

Over to you

The Times, 14 December 2004

According to a newspaper article, research has shown that inability to identify ten particular smells is an early sign of Alzheimer's disease. What do you think the ten smells are?

27 The nervous system 2

A The motor system

Examination of the motor system should include assessment of the following:

- **muscle bulk** (amount of muscle tissue). Look for signs of **wasting** (muscle atrophy)
- **muscle tone** (amount of tension in a muscle when it is relaxed). Tone can be increased (**spasticity**), or decreased (**flaccidity**)
- **muscle power** (strength)
- **coordination** (the ability to use several muscles at the same time to perform complex actions)
- **gait** (the manner of walking)
- **reflexes** (see B below)
- **involuntary** movements, for example a **tic** or a **tremor**.

Here is an extract from a case report about a patient with a tremor.

Case 80

On examination, her face showed little or no **expression**. There was a tremor **affecting** mainly her right hand. She had **generally increased** muscle tone. Power, reflexes, coordination and sensation were **within normal limits**. Examination of her gait showed that she was slow to start walking and had difficulty stopping and turning.

B Tendon reflexes

Examination of the nervous system normally includes testing the **tendon reflexes**, for example the **knee jerks**, with a **tendon hammer** (also known as a **reflex hammer**). The reflexes may be **absent** (0), **diminished** (–), **normal** (+) or **brisk** (+++). The **plantar reflexes** are also checked. The normal plantar response is a **downgoing** (↓) movement (plantar flexion) of the big toe. An **upgoing** (↑) toe (extensor or **Babinski** response) is abnormal.

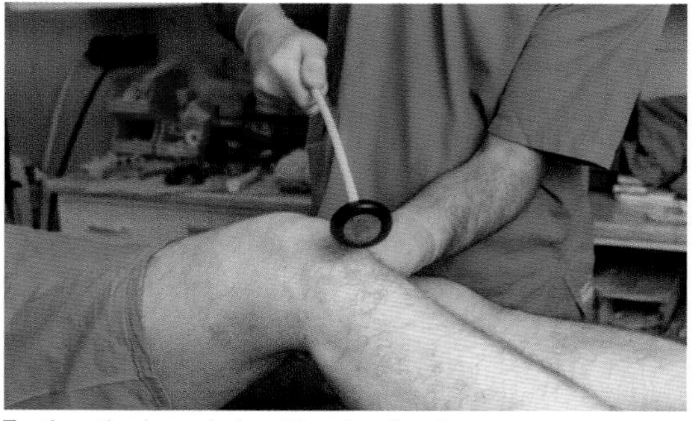

Testing the knee jerk with a tendon hammer

C Coma

Coma is unconsciousness with a reduced response to external stimuli.

Doctors say: "The patient is **in a coma.**" "The patient is **comatose.**"

The **Glasgow Coma Scale (GCS)** score is calculated as follows:

Eye opening		Verbal response		Motor response	
Spontaneous	4	Oriented	5	Obeys	6
To speech	3	Confused	4	Localizes	5
To pain	2	Inappropriate	3	Withdraws	4
None	1	Incomprehensible	2	Flexion	3
		None	1	Extension	2
				None	1

27.1 Complete the table with words from A and B opposite.

Noun	Adjective
absence	
diminution	
	flaccid
	spastic
	wasted

27.2 A doctor is giving instructions to a patient during examination of the motor system. Identify what the doctor is assessing in each case. Look at A opposite to help you.

1 I'd like you to relax. I'm just going to move your arm up and down.
2 Can I see your hands?
3 Now, I'm going to straighten your arm out. Try to stop me.
4 Can you touch my finger with yours and then touch your nose? Good. Now do it again with your eyes closed.

27.3 Complete the sentences. Look at A, B and C opposite and at the table in 27.1 above to help you.

1 A .... hand droops limply to form a right angle with the wrist.
2 .... reflexes are reflexes that are stronger than normal.
3 Muscle .... means the muscle is reduced in bulk.
4 A tic is a form of .... movement.
5 A key is often used to test the .... response.
6 His .... was poor: he could not perform rapid alternating movements.
7 A .... is used to test reflexes.
8 When something is .... , it is less than normal.

27.4 A patient is brought to A&E in coma. When her name is spoken, she opens her eyes but she does not answer questions, or obey instructions. What is her GCS score?

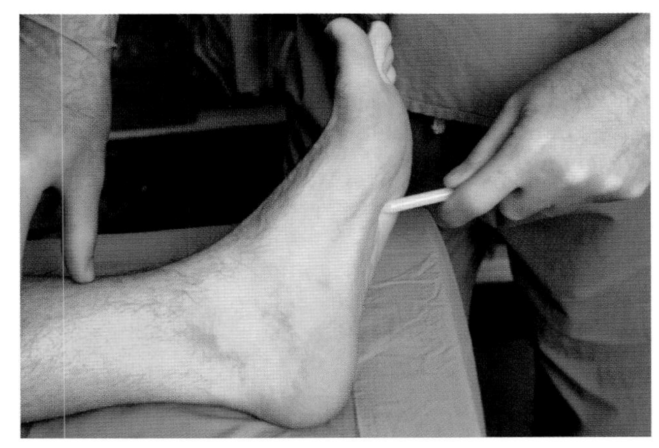

The plantar reflex

Over to you

Can you name six tendon reflexes?
What is your diagnosis for the patient in A opposite?

28 Oncology

A Neoplasms

A **neoplasm** is an abnormal new growth of tissue. **Malignant** neoplasms – cancers – are likely to spread and cause serious illness or death. **Benign** neoplasms do not spread and are less harmful.

When speaking to patients, doctors generally say **growth** or **tumour**.

You have a small	growth tumour	in the bowel.

A **lump** or **swelling** is a collection of tissue or fluid which is visible or **palpable** – can be palpated or felt with the fingers. A lump may be due to a neoplasm, but there are other causes such as inflammation or fluid accumulation.

Patients say:

> I have a lump in my left breast.

Doctors say:

> There was a firm, palpable **mass** in the liver.

Malignant tumours are characterized by rapid growth and **invasiveness.** The tumour may **invade** local tissues or may **spread to** distant parts of the body (**metastasis**). Neoplasms which are the result of metastasis are called **secondaries**, as opposed to the original tumour which is the **primary**.

> We have the results of the scan back now and I'm afraid they show that you have a small growth in the prostate. Fortunately, it appears to be **at an early stage** and there is **no sign of spread**. So if we **remove** it, there is every chance of a complete **cure**.

Note: To **invade** (verb) is to enter and spread throughout a part of the body, and this process is **invasion** (noun). If a tumour is described as **invasive**, it has the ability to spread.

B Symptoms and signs of malignancy

The symptoms of malignant disease may be related to the size and location of the tumour. For example, a **space-occupying lesion** in the brain causes raised intracranial pressure and symptoms such as headache, vomiting, or visual disturbance. Tumours of the colon may **obstruct** (**block**) the lumen and cause change in bowel habit. Other possible symptoms of malignancy include bleeding, pain, and weight loss.

Case 55

A 33-year-old man presented to his GP complaining of a **painless lump** on the right side of his neck, which had been **present** for about two months and was **enlarging**. He had been feeling generally unwell and had lost about 5 kg in weight. He was also complaining of night sweats. He had no significant past medical history.

C Treatment of tumours

A tumour can sometimes be completely removed or **excised** by surgery. If this is not possible, for example if it has already metastasized to other parts of the body, it may be possible to destroy it by radiotherapy or by chemotherapy (see Unit 42). When a cure is not possible, **palliative treatment** is given, which is only intended to relieve symptoms.

28.1 Complete the table with words from A, B and C opposite and related forms.

Verb	Noun(s)	Adjective(s)
cure		curative
	excision	
grow		growing
		invasive
	obstruction	obstructive
palliate	palliation	
		palpable
	spread	spreading
swell		swelling, swollen

28.2 The notes below are about the patient described in B opposite. Use them to put the sentences (1–9) in the correct order, to make the next paragraph of the case report. Use Appendix II on page XX if you need help with the abbreviations.

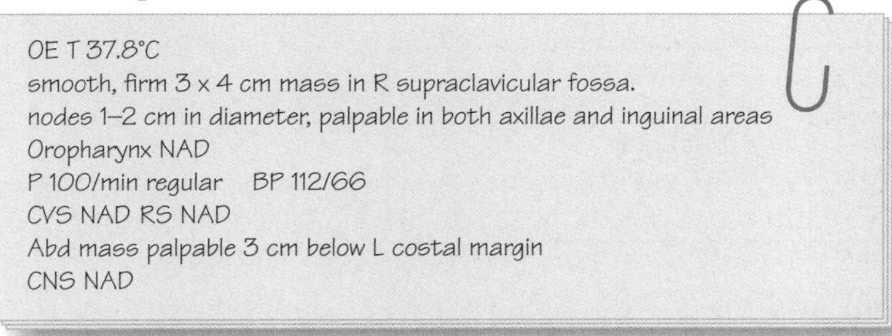
OE T 37.8°C
smooth, firm 3 x 4 cm mass in R supraclavicular fossa.
nodes 1–2 cm in diameter, palpable in both axillae and inguinal areas
Oropharynx NAD
P 100/min regular BP 112/66
CVS NAD RS NAD
Abd mass palpable 3 cm below L costal margin
CNS NAD

1 Examination of the central nervous system was normal.
2 His mouth and throat were normal.
3 There was a smooth, firm 3 x 4 cm mass in the right supraclavicular fossa.
4 His cardiovascular and respiratory systems were normal.
5 On abdominal examination, there was a mass palpable 3 cm below the left costal margin.
6 On examination, his temperature was 37.8°C.
7 There were enlarged lymph nodes in both axillae and inguinal areas.
8 His pulse rate was 100/min regular and blood pressure 112/66.

28.3 Complete the sentences. Look at A, B and C opposite to help you.

1 Distant .... of tumour cells is known as metastasis.
2 Many symptoms of cancer, such as difficulty swallowing, are due to ....
3 The opposite of painful is ....
4 A .... neoplasm is called a cancer.
5 Tumours which do not invade or metastasize are ....
6 .... tumours are those which result from the spread of a primary.
7 If a cure is not possible, ... treatment should be given.
8 A .... liver suggests metastasis.

Over to you

The most likely clinical diagnosis in the patient (described in B opposite and 28.2 above) is lymphoma. How would you explain his condition to him?

29 Pregnancy and childbirth

A Childbirth

The **expected date of delivery** (EDD) is the date on which a woman is expected to give birth to the child she is **carrying** (pregnant with). It is calculated by adding 280 days or 40 weeks to the first day of the last menstrual period (LMP). **Childbirth** is also referred to by doctors as parturition. **Delivery** is the process of helping the child to be born. A **spontaneous vaginal delivery** (SVD) is a normal delivery. If there are complications, the baby may be **delivered** by **caesarean section** (surgically removed).

A **full-term pregnancy** is 40 weeks, divided into three **trimesters.** A baby who is born before this is **premature**, and one born after 40 weeks is **postmature.** A baby who is born dead, for example because the **umbilical cord** is around its neck, is **stillborn.** A pregnancy may end before term spontaneously, with a **miscarriage** (**spontaneous abortion**), or be deliberately **terminated** with an **induced abortion** (**termination of pregnancy**).

Note: the verb **induce** means to cause something to happen.

B Labour

The process by which the fetus and placenta are pushed out of the uterus is called **labour.** It is divided into four stages. Some words which are combined with labour are:

premature **prolonged** **spontaneous** **induced** **false**	labour

C Presentation and lie

Fetal lie is the position of the fetus in the uterus. The normal lie is **longitudinal**, and the abnormal lie is **transverse. Fetal presentation** refers to 'the part of the fetus which occupies the centre of the pelvic canal and which the examining finger feels on vaginal examination' (Butterworth). The normal presentation is with the head (**vertex presentation**). **Breech presentation** means the buttocks are presenting (*breech* is an old word for buttocks). Abnormal presentations may require delivery with **forceps.**

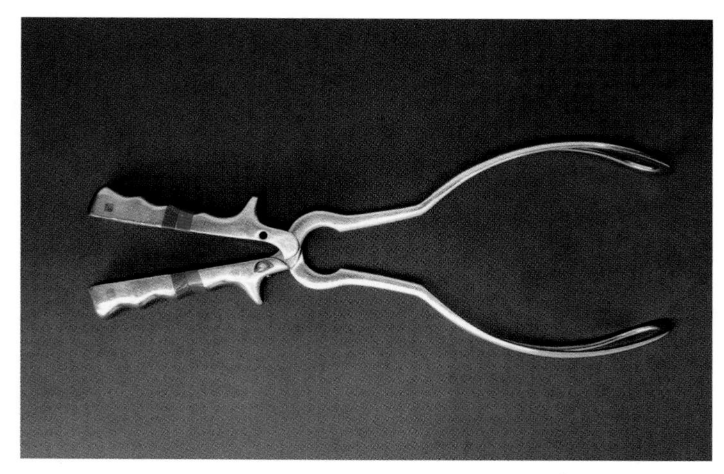

Obstetric forceps

29.1 Complete the sentences. Look at A, B and C opposite to help you.

1 A baby that is born a week before the EDD is ....
2 A .... of pregnancy may be necessary for medical reasons.
3 The first three months of pregnancy are known as the first ....
4 Fetal distress in the first stage of .... is an indication for caesarean ....
5 It was a breech .... and delivery was by forceps.
6 A .... is another term for a spontaneous abortion.
7 The .... was wound tightly around the baby's neck and it was unfortunately ....

29.2 Complete the table with words from A opposite.

Verb	Noun
abort	
deliver	
	induction
miscarry	
present	
	termination

29.3 Dr Bennett, an SHO, is presenting a patient at a weekly meeting in the obstetric unit of a hospital. Complete the presentation with the correct forms of verbs from 29.2 above.

This is Clara Davis. She came to the antenatal clinic at nine weeks. In her past obstetric history, she had a pregnancy when she was 18, which was (1) .... and another one a year later, which spontaneously (2) .... Since then she has had three pregnancies. In the first, the baby was (3) .... normally at 40 weeks. In the second, she had an (4) .... of labour at 39 weeks because of fetal distress. The third baby (5) .... as a breech and was (6) .... by caesarean section.

Over to you

Romanian woman gives birth at 66

A 66-year-old woman is believed to have become the world's oldest mother yesterday, after giving birth to a girl.

The Guardian, 17 January 2005

What is the oldest and youngest age for giving birth that you have known? In your opinion, should assisted conception be available for anyone who wants it? If not, what do you think the limits should be?

30 The respiratory system

A Cough

Cough is a common symptom of **upper respiratory tract infection** (URTI) and lung disease. A cough may be **productive**, where the patient coughs up **sputum**, or **non-productive**, where there is no sputum. A productive cough is often described as **loose** and a non-productive cough as **dry**. Sputum (or **phlegm**) may be clear or white (**mucoid**), yellow due to the presence of pus (**purulent**), or **blood-stained** (as in haemoptysis).

A doctor is examining a patient who is complaining of a cough.

Doctor: How long have you had the cough?
Mr Hamilton: Oh, for years.
Doctor: Do you smoke?
Mr Hamilton: I used to **smoke heavily**, but I **gave up** a year ago.
Doctor: Do you **cough up** any **phlegm**?
Mr Hamilton: Yes.
Doctor: What colour is it?
Mr Hamilton: Usually yellow.
Doctor: Have you ever **noticed any blood** in it?
Mr Hamilton: No.
Doctor: Any **problems with your breathing**?
Mr Hamilton: Yes, I get very short of breath. I have to stop halfway up the stairs to **get my breath back**.

The doctor writes in the patient's case notes:

c/o dyspnoea & cough c. purulent sputum for years. No haemoptysis.

Note: The noun **phlegm** is pronounced /flem/.

B Auscultation

The doctor is examining Mr Hamilton's chest.

Take deep breaths in and out through your mouth. Good. Now **say 'ninety-nine'**.

Listening to the chest with a stethoscope may reveal the presence of sounds, apart from the normal **breath sounds**. There are two main kinds of **added sounds**:

- **crackles**, which sound like hairs being rubbed together and suggest the presence of fluid in the lungs
- **wheezes**, which are more musical sounds, like whistling, and indicate narrowing of the airways. The sound of an asthma patient's breathing is also called wheeze.

The sound heard when the pleural surfaces are inflamed, as in pleurisy, is called a **pleural rub**.

The doctor asks Mr Hamilton to say 'ninety-nine' to check **vocal resonance**, which may be increased (as in pneumonia), or decreased (as in pneumothorax).

After examining Mr Hamilton, the doctor adds to his notes:

OE Chest: early inspiratory crackles both lung bases + expiratory wheeze

30.1 Make word combinations using a word from each box. Look at A and B opposite to help you.

blood- breath pleural productive vocal	cough rub stained resonance sounds

30.2 Rewrite the questions, using words that are better known to patients. Look at A opposite to help you.

1 Is your cough productive?
2 What colour is the sputum?
3 Is it ever purulent?
4 Have you ever had haemoptysis?
5 Do you suffer from dyspnoea?

30.3 Are the following statements true or false? Give reasons for your answers, using your medical knowledge and A and B opposite to help you.

1 A patient who has a loose cough produces phlegm.
2 Crackles are heard when the airways are narrowed.
3 A patient who has a non-productive cough produces sputum.
4 Wheezes are typical of pleurisy.
5 A pleural rub is a sign of asthma.

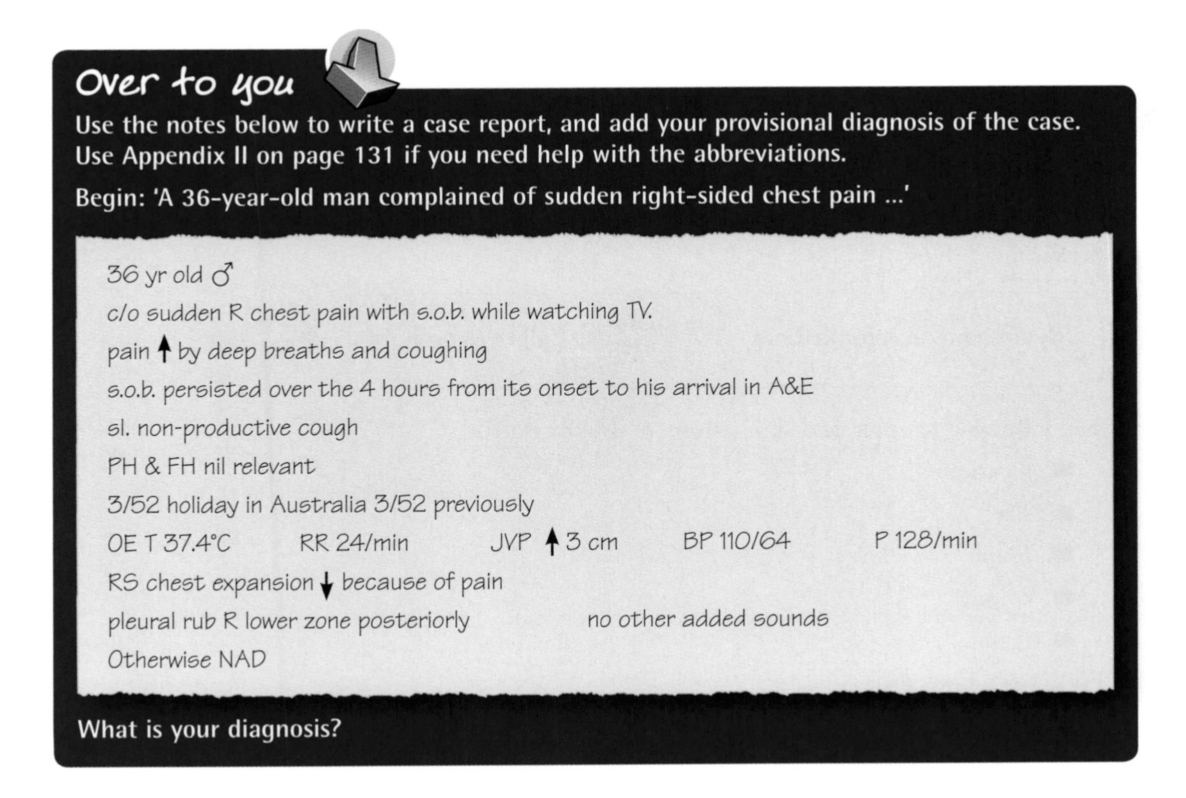

Over to you

Use the notes below to write a case report, and add your provisional diagnosis of the case. Use Appendix II on page 131 if you need help with the abbreviations.

Begin: 'A 36-year-old man complained of sudden right-sided chest pain ...'

36 yr old ♂
c/o sudden R chest pain with s.o.b. while watching TV.
pain ↑ by deep breaths and coughing
s.o.b. persisted over the 4 hours from its onset to his arrival in A&E
sl. non-productive cough
PH & FH nil relevant
3/52 holiday in Australia 3/52 previously
OE T 37.4°C RR 24/min JVP ↑ 3 cm BP 110/64 P 128/min
RS chest expansion ↓ because of pain
pleural rub R lower zone posteriorly no other added sounds
Otherwise NAD

What is your diagnosis?

31 The skin 1

A Some types of skin lesion

Medical term	Common word	Features
macule	**spot**	not raised above the surface of the skin
papule	**spot**	raised above the surface of the skin
nodule	**lump**	a large papule
vesicle	**small blister**	filled with fluid
bulla	**blister**	a large vesicle
pustule	–	filled with pus
crust	**scab**	dried blood etc. on the surface of the skin
scales	**scales**	a thin layer of epidermis separated from the skin
cicatrix (plural: cicatrices)	**scar**	a mark on the skin after healing
naevus	**birthmark**	a coloured skin lesion present at birth
fleshy naevus	**mole**	a raised brown naevus
verruca	**wart**	a nodule produced by HPV
furuncle	**boil**	a large pustule, or skin abscess

Note: The liquid (often yellow) formed as a result of infection is **pus**. If a lesion is **pustular**, it is filled with pus.

B Rashes

A **single** skin **lesion** can be **regular** or **irregular** in shape. When there are many (**multiple**) **lesions**, especially macules or papules, the result is a **rash**, (or **spots** in common language); for example the rash of an infectious disease such as rubella. A rash is said to **erupt**, or **break out**.

My little boy has **broken out**	in spots in a rash	all over his body.

The following features of a skin lesion are usually noted:

- location
- size
- shape
- colour
- type.

For a rash, note also:

- **distribution** (**widespread** – on many parts of the body, or **localized** – on one part only)
- **grouping** (**scattered** – more or less evenly spread out, or in **clusters** – small groups).

31.1 Complete the description of *herpes zoster* (shingles) by replacing the medical words in brackets with ordinary English words. Look at A and B opposite to help you.

(1) .... (herpes zoster) usually starts with pain and soreness. Then red (2) .... (macules) appear that develop into groups of (3) .... (vesicles) over a particular area on one side of the body. In most patients, new (4) .... (lesions) continue to appear for 3 to 5 days. The (5) .... (vesicles) become (6) .... (pustular) and then form (7) .... (crusts). In severe cases, there may be (8) .... (cicatrices) afterwards.

(*BMJ* 2005; 331: 148 Amended with permission from the BMJ Publishing Group)

31.2 Read the description of the rash of rubella and complete the notes. Look at A and B opposite to help you.

The spots are scattered pink macules which appear first behind the ears and on the forehead. The rash spreads rapidly, first to the trunk and then to the limbs.

location and distribution:

grouping:

type of lesion:

colour:

31.3 Complete the notes for the rash in the photograph, and suggest a diagnosis. Look at A and B opposite to help you.

location and distribution:

grouping:

type of lesion:

colour:

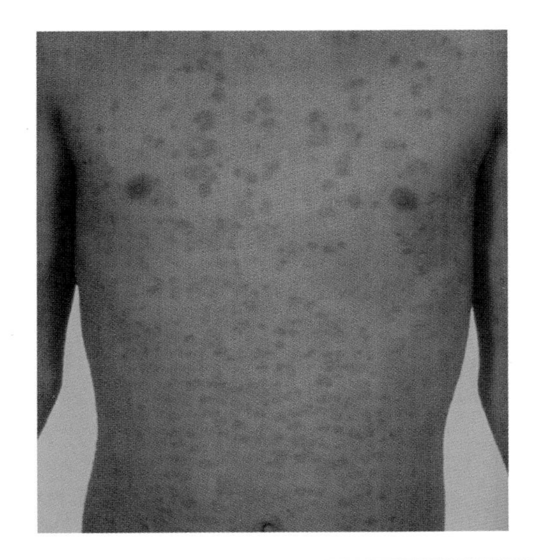

31.4 Complete the notes for the lesion in the photograph, and suggest a diagnosis. Look at A and B opposite to help you.

location and distribution:

grouping:

type of lesion:

colour:

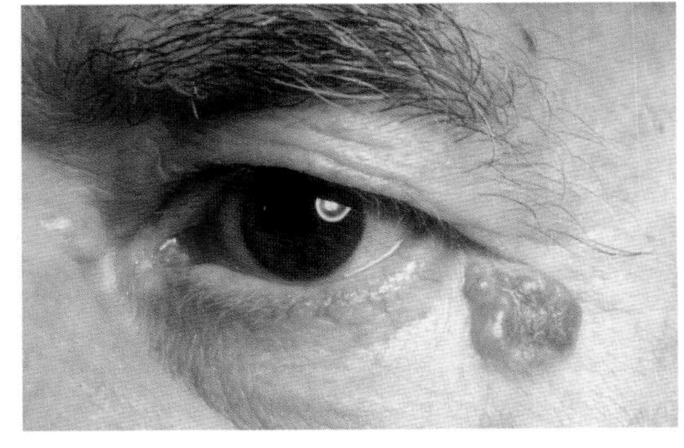

Over to you

What types of rashes are common in your country? Is there any reason why they are common?

32 The skin 2

A Injuries to the skin

Here is an extract from a medical website.

Mechanical injuries to the skin are divided into those caused by a **blunt force**, such as a punch from a fist, and those caused by a **sharp force**, such as a knife.

Injuries from blunt forces

- An **abrasion** (also called a **graze** or a **scratch**) is a **superficial** (surface) injury involving only the epidermis, which has been removed by friction. A scratch is linear, as in fingernail scratches, whereas a graze involves a wider area, as in abrasions caused by dragging part of the body over a rough surface.
- A **contusion** (also called a **bruise**) is an injury that occurs when blood vessels in the skin are damaged.
- A **laceration** (also called a **tear**) is a wound involving both the dermis and epidermis. It is usually distinguished from penetrating or incised wounds by its irregular edges and relative lack of bleeding.

Injuries from sharp forces

- An **incised wound** (also called a **cut**) is a break in the skin where the length of the wound on the surface is greater than the depth of the wound – for example, a wound caused by a razor blade.
- The depth of a **penetrating wound** is greater than the superficial length of the wound – for example, a **stab wound** caused by a knife.

(Amended with permission from the BMJ Publishing Group)

B Case report

Read the case report and compare it with the illustration.

Case 2

A 9-year-old boy presented to the Accident and Emergency department after he stumbled and fell while running in a wood. He had received a **blow** to the head from a rock and had been **scratched** by bushes. On examination, a vertical laceration 1 cm long was noted on the bridge of his nose just right of the midline. There were a number of superficial scratches on the right side of his forehead. His right upper lid was mildly **contused**.

(BMJ 1998; 316: 1364
Amended with permission from the BMJ Publishing Group)

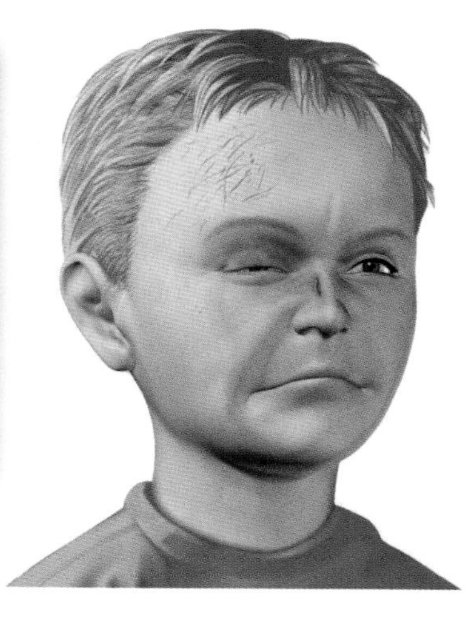

C Sores

The word **sore** is a popular term for many different types of skin lesion, especially infected lesions. A **pressure sore** is a skin ulcer caused by pressure, for example the pressure of lying in bed for long periods (also known as a **bedsore**, or decubitus ulcer). A **cold sore** is a lesion caused by *herpes simplex*.

Note: The adjective **sore** means painful, for example a **sore throat**.

32.1 Write the corresponding medical terms for the ordinary English words and say what kind of force is involved. Look at A opposite to help you.

Common word	Medical term	Type of force
bruise		
cut		
graze		
scratch		
stab wound		
tear		

32.2 Choose the correct words to complete the description of the injuries shown in the illustration. Look at A and B opposite to help you.

There are (1) .... (scratches/grazes) above the left eyebrow and on the left side of the neck, a (2) .... (contusion/laceration) to the left side of the lower lip and (3) .... (cuts/tears) to the left cheek.

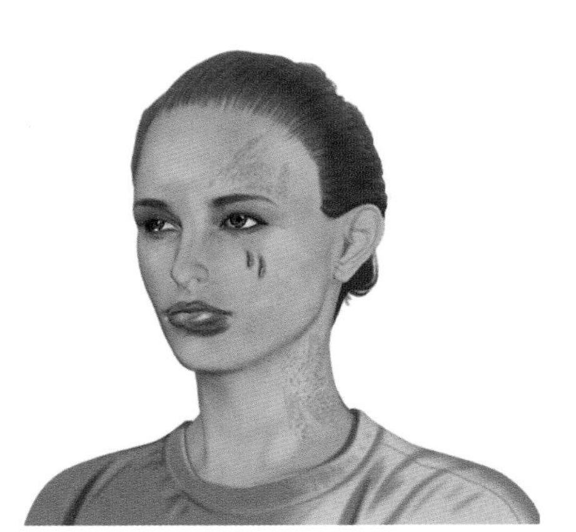

32.3 Write a description of the injuries shown in the illustration. Look at A and B opposite and at 32.2 above to help you.

....

....

....

....

....

....

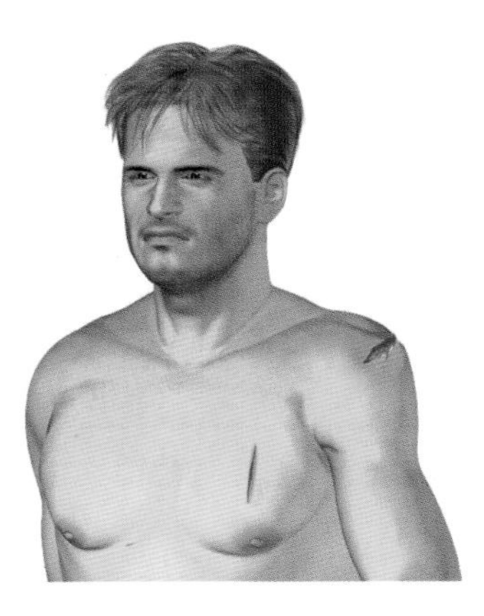

32.4 Complete the sentences. Look at A, B and C opposite to help you.

1 Frequent changes of position are necessary in the immobile patient to prevent the development of a pressure ....
2 He had several .... wounds in the abdomen from the knife.
3 He was knocked unconscious by a heavy .... to the head.
4 The wounds were only .... and required no treatment.

Over to you

The police have asked you to examine a man who has been involved in a fight in a restaurant. What type of injuries would you expect to find, and how might they have been caused?

33 The urinary system

A Urinary symptoms

Urine is formed in the kidneys and stored in the bladder until it is **passed** (or **voided**).

Patients may say:

I have some pain when I **pass water.** / **pee.**

Doctors may say:

Are you having any trouble with your **waterworks?**

Doctor: Are you having any trouble with your waterworks?
Mr Jones: Well, I do seem to have to **go to the toilet** more often than I used to.
Doctor: How often is that?
Mr Jones: It depends, but sometimes it's every hour or even more often.
Doctor: What about at night? Do you have to get up at night?
Mr Jones: Yes. Nearly always two or three times.
Doctor: Do you get any burning or pain when you pass water?
Mr Jones: No, not usually.
Doctor: Do you have any trouble getting started?
Mr Jones: No.
Doctor: Is the **stream** normal? I mean is there still a good strong **flow**?
Mr Jones: Perhaps not quite so good as it used to be.
Doctor: Do you ever **lose control of your bladder**? Any **leaking** or **dribbling**?
Mr Jones: Well, perhaps a little dribbling from time to time.
Doctor: Have you ever **passed blood** in the urine?
Mr Jones: No, never.

Common urinary symptoms and their definitions:

frequency	frequent passing of urine
dysuria	**burning** or **scalding** pain in the urethra when passing urine
nocturia	urination at night
urgency	urgent need to pass urine
hesitancy	difficulty starting to pass urine
urinary incontinence	involuntary passing of urine
haematuria	macroscopic blood in the urine

B Urinalysis

Urinalysis is the analysis of urine. Simple screening tests of the urine are carried out with **reagent strips**, for example Clinistix for the detection of glucose. More detailed tests are carried out in a laboratory on a **specimen** of urine. Typical specimens are a **midstream specimen** (**MSU**) and a **catheter specimen** (**CSU**). Microscopic examination may reveal the presence of red blood cells, pus cells, or casts. **Casts** are solid bodies formed by protein or cells.

Plus signs are used in case notes to indicate abnormal findings. A small amount (+) is described as a **trace**. For a large amount (+++), the words **gross** or **marked** can be used, for example **gross haematuria**. When there is nothing, the word **nil** is common.

sugar nil
protein +
blood +++

There was no sugar, a **trace** of protein and **gross** haematuria.

33.1 Look at the conversation in A opposite and complete the notes about Mr Jones. Use medical terms where possible.

c/o (1) .... and (2) .... for 1 yr.
No (3) .... or (4) ....
(5) .... a little weaker.
No incontinence apart from occasional (6) ....

33.2 Match the patients' descriptions of their symptoms (1–7) with the medical terms (a–g). Look at A opposite to help you.

1 'I have to pee every half hour or so.'
2 'I get a scalding pain when I pass water.'
3 'I have to get up several times to pass water at night.'
4 'I have to rush to go to the toilet.'
5 'I have trouble getting started.'
6 'I can't hold my water.'
7 'I passed some blood in my urine.'

a dysuria
b frequency
c haematuria
d hesitancy
e nocturia
f urgency
g incontinence

33.3 Write the doctor's questions for each of the symptoms in 33.2 above. Look at A opposite to help you. You will need to think of your own question for urgency.

33.4 Describe the findings of the laboratory report in words. Look at B opposite to help you.

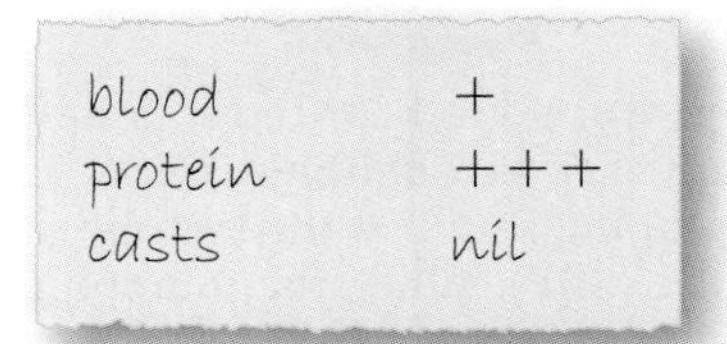

blood	+
protein	+++
casts	nil

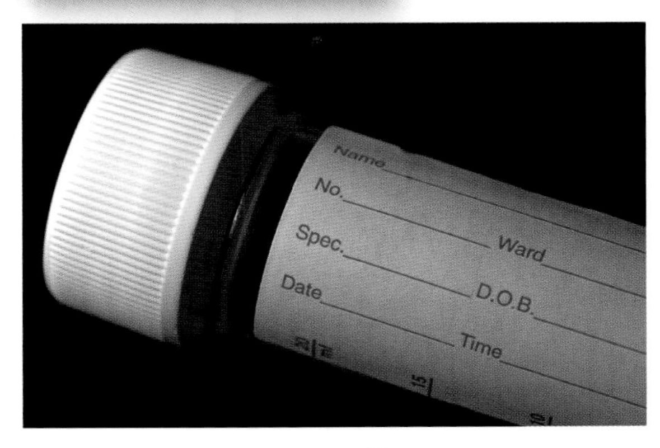

Over to you

What is your provisional diagnosis for Mr Jones?

34 Basic investigations

A Ophthalmoscopy

An ophthalmoscope allows the doctor to examine all parts of the eye: the iris, lens, retina and optic disc. For best results, the examination is done with **dimmed**, or lowered, lights to allow the **pupil** to maximally **dilate** or widen. A **topical mydriatic solution** may be applied to the eye to aid **dilation**. The patient is then asked to **fixate** on a target for the duration of the test.

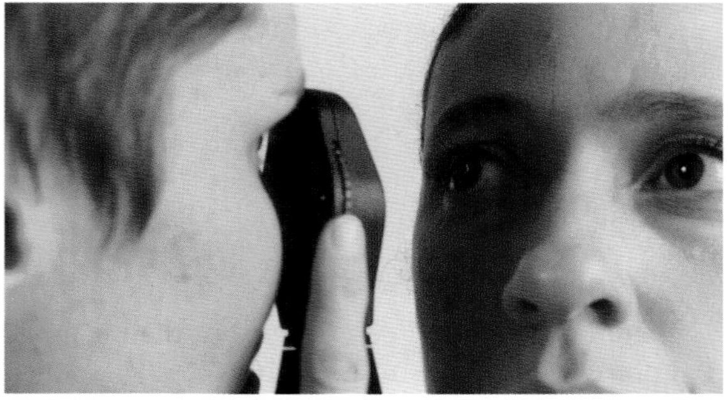

Direct ophthalmoscopy

B Blood pressure

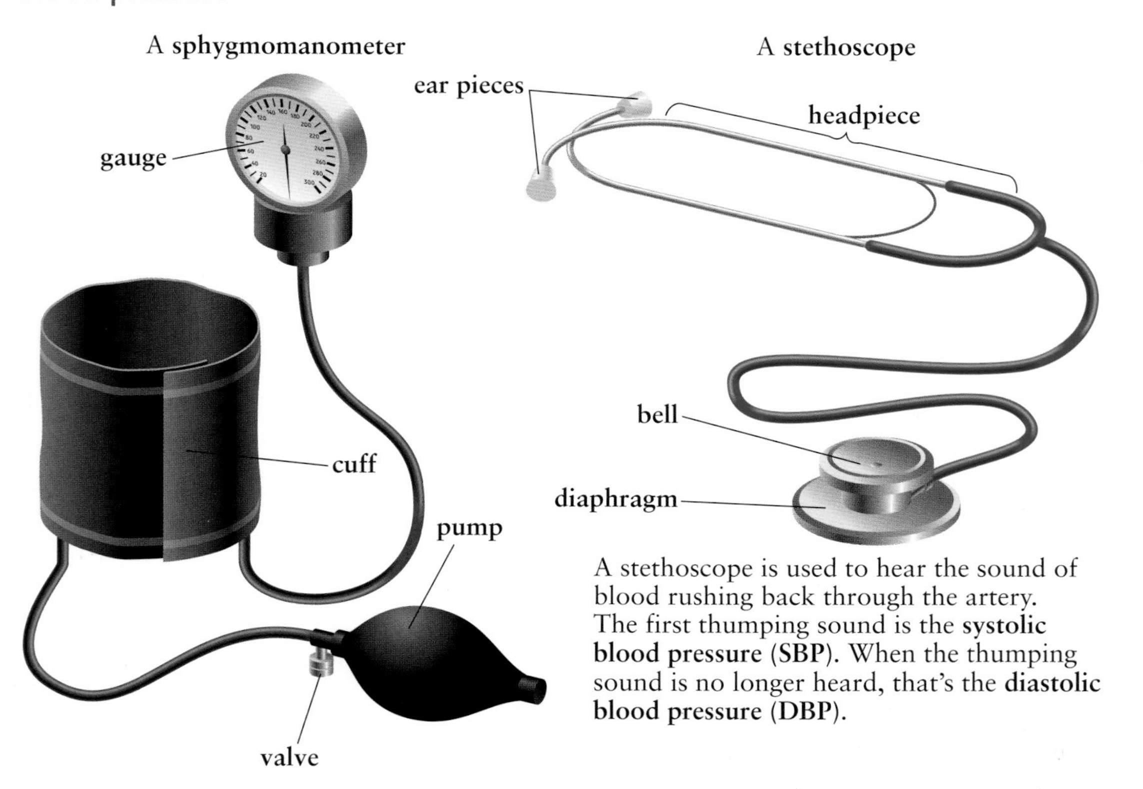

A stethoscope is used to hear the sound of blood rushing back through the artery. The first thumping sound is the **systolic blood pressure (SBP)**. When the thumping sound is no longer heard, that's the **diastolic blood pressure (DBP)**.

C Taking blood

During **venipuncture**, the **phlebotomist**, a technician who takes blood, inserts a needle into a vein and withdraws blood into a **specimen tube**, which is sent to the haematology **laboratory** for **analysis**. Usually the phlebotomist can find a vein in the inner part of the elbow, the antecubital fossa, that is easily accessible. She may **apply** a **tourniquet** – a tight band – above the site, or the patient may be asked to clench their hand to **make a fist**, in order to make the vein more prominent. Afterwards, the patient may be asked to press lightly on a **dressing**, usually a piece of gauze, to help the blood to **clot** and to prevent swelling and a haematoma (a black and blue mark, or a **bruise**) where the vein was punctured.

Note: A **bruise** is a specific mark. **Bruising** can be used to describe a number of bruises or a larger area – *The patient exhibited bruising on the right forearm.*

34.1

A doctor is talking to a patient during an ophthalmoscopy. Match the underlined expressions she uses (1–4) to expressions with similar meanings from A opposite (a–d).

> Right, Mr Gold, because you've been having these headaches I'm going to have a look at your eyes, particularly the back of your eye – the retina. I'm going to put (1) a couple of drops in your eye, (2) which will make it easier for me to see the retina. After a few minutes you may find your vision a bit blurry. This will wear off after about an hour.
>
> (3) I need to get rid of as much external light as possible. This means closing the blinds.
>
> Now, I'd like you to (4) look straight ahead at that clock. This takes a few minutes and your eyes might feel a bit tired so you can blink if you need to. I don't want you to look at me, look at the clock.

a the examination is done with dimmed lights
b a topical mydriatic solution
c to aid dilation
d fixate on a target

34.2

Complete the instructions. Look at B opposite to help you.

1 Wrap the .... around the patient's upper arm.
2 Place the .... over the area of the brachial artery. Raise the patient's arm so that the brachial artery is at the same height as the heart.
3 Close the valve on the ....
4 Pump up the pressure to at least 150 mmHg. Open the .... a little and slowly deflate the cuff while listening and watching the pressure ....
5 The first sound you hear is the flow of blood through the brachial artery. The value on the gauge at that point is the ....
6 Continue listening while you slowly .... the cuff.
7 The .... blood pressure is measured when the sound completely disappears.

34.3

Complete the text. Look at C opposite to help you.

(1) .... are specially trained in taking blood. They are skilled at (2) .... – puncturing the vein to take a blood sample. The wrist, hand and foot can be used but more often a vein in the inner part of the elbow is used. If it is difficult to locate a suitable vein, the patient may be asked to make a (3) ...., or a (4) .... may be applied on the upper arm to make the vein more apparent. Afterwards, a (5) .... is applied and the patient is asked to press gently. This helps to stop the bleeding and prevent (6) .... at the site. It is important that (7) .... are labelled correctly before they are sent to the haematology (8) ...., where a full blood count or other investigations will be carried out.

Over to you

Practise talking a patient through an investigation that you carry out regularly.

35 Laboratory tests

A A Microbiology request form

A Microbiology request form uses a number of abbreviations for specimen types (see Appendix II on page 131 for an explanation of these abbreviations).

Date collected .../.../...
Time collected .... (24hr)

- ☐ **MSU**
- ☐ **CSU**
- ☐ Urine – Other, specify
- ☐ Urine first voided – for chlamydia
- ☐ Faeces
- ☐ Sputum
- ☐ **NP** secretions
- ☐ **BAL**
- ☐ Induced sputum

- ☐ Nose **sw**
- ☐ Throat sw
- ☐ Axilla sw
- ☐ Groin sw
- ☐ Eye sw
- ☐ Endocervical sw
- ☐ **Sw in Virus TM*** (give site)
- ☐ Sw for chlamydia (give site)

*special medium

- ☐ **Blood Culture**
- ☐ Clotted Blood
- ☐ **EDTA blood**
- ☐ **CSF**

Other:

B A Biochemistry and Haematology lab report

	Value	Range	Unit
Full blood count (FBC)			
Haemoglobin (Hb)	143	115–165	g/L
Haematocrit (HCT)	0.224	0.37–0.47	L/L
Mean cell volume (MCV)	72.5	78.0–98.0	fL
White cell count (WCC)	7.4	4.0–11.0	10^9/L

	Value	Range	Unit
Urea and electrolytes (U&E)			
Urea	4.5	2.5–6.6	mmol/L
Creatinine	58	60–120	umol/L
Sodium (Na)	138	135–145	mmol/L
Potassium (K)	4.5	3.6–5	mmol/L
Liver function test (LFT)			
Bilirubin	7	3–16	umol/L
ALT	9	10–50	U/L
Alkaline Phosphatase (Alk.Phos)	131	40–125	U/L

C Terms used to describe lab results

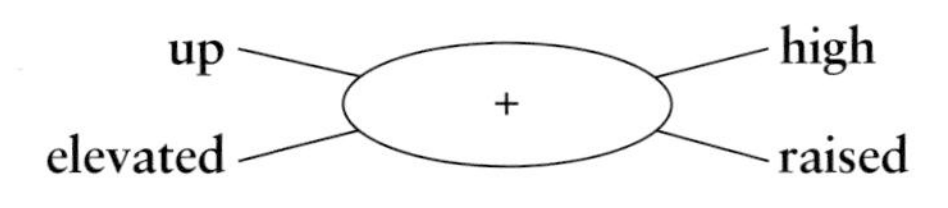

down – low – reduced

When the results are within the normal range, doctors say:

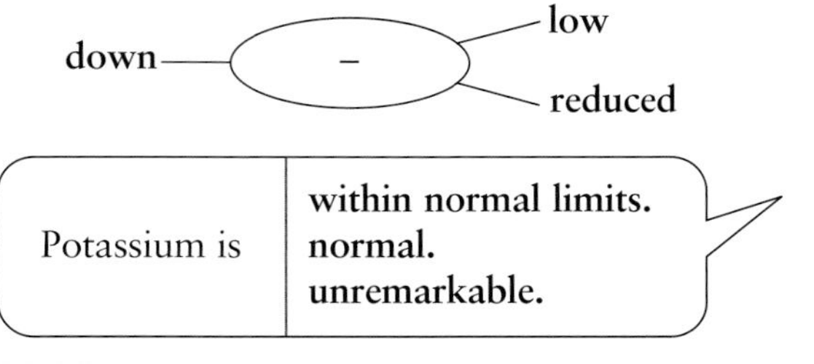

Unit abbreviation	Full form
g/L	grams per litre
L/L	litres per litre
10^9/L	times ten to the power nine per litre
fL	femtolitres
mmol/L	millimols per litre
umol/L *or* µmol/L	micromols per litre
U/L	units per litre

35.1 Write the name of the specimen for each of the suspected conditions. Look at A opposite to help you.

Suspected condition	Specimen
anaemia	
bacterial conjunctivitis	
genital herpes	
meningitis	
septicaemia	
urinary infection	
urinary infection (catheter in place)	

35.2 Complete the sentences describing the results of the report in B opposite. Look at C opposite to help you.

1 Haemoglobin is ...., one hundred and forty-three .... per litre.
2 Creatinine is slightly .... , fifty-eight .... litre.
3 Alkaline Phosphatase is .... , one hundred and thirty-one ....
4 ALT is slightly reduced, nine ....
5 Bilirubin is .... , seven ....

35.3 Write full descriptions of the following results from a case history. Look at B and C opposite and at 35.2 above to help you.

Na 138, K 4.5, WCC 12.2, HCT 0.224, MCV 72.5, Alk.Phos 72, ALT 9

Sodium is normal, one hundred and thirty-eight millimols per litre.

Over to you

Describe the lab results of a recent case you've been involved in.

36 Endoscopy

A Functions of endoscopy

Endoscopy is a way of examining parts of the body which are not visible from the outside. A typical endoscope is a flexible **tube** which is **inserted through** one of the natural **orifices** – openings – such as the anus or mouth. **Rigid endoscopes**, which cannot be bent, are also used but are inserted through small **incisions** – surgical cuts. The **shaft** contains several **channels** to transmit light from the outside and images from inside and to allow different instruments to be used.

Endoscopes can be used for the following:

- to provide diagnostic information
- to **excise** – cut out – diseased tissue or **growths** such as **polyps**
- to clear obstructions
- to **take a biopsy**
- to **cauterize** a site of bleeding by applying heat.

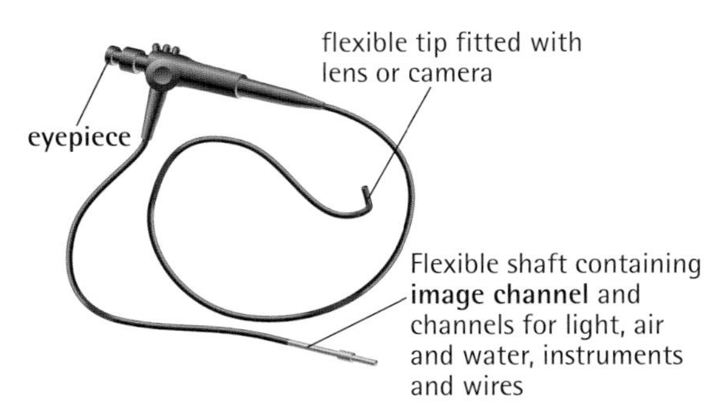

B Enteroscopy

Dr Jardine is talking her patient through an enteroscopy.

Now, I'm just **lubricating** the tube with a **jelly** which contains a **local anaesthetic**. It'll help to ensure a smooth passage as it **passes down** and you shouldn't feel too much.

I'm going to **feed** the tube **through** your nose. This is the most uncomfortable part of the procedure but it's very brief. You'll **get used to** the tube in a few minutes' time. OK, when it hits the back of your throat, **take a** deliberate **swallow**. I'll tell you when.

Now! **Swallow**, swallow. That's it. Well done.

C Report of a diagnostic endoscopy

EXAMINATION

Informed consent was **obtained** from the patient after discussing **risks and benefits** of the procedure. The patient was connected to the **pulse oximeter** and placed **in the left lateral position**. Oxygen was provided through a **nasal cannula** and the **premedication administered** as stated. The endoscope was **introduced into** the oesophagus. At the end of the examination the patient was **transferred to** the **recovery area** to **recuperate**.

PREMEDICATION	ENDOSCOPE
Throat spray	Olympus GIF-XQ240

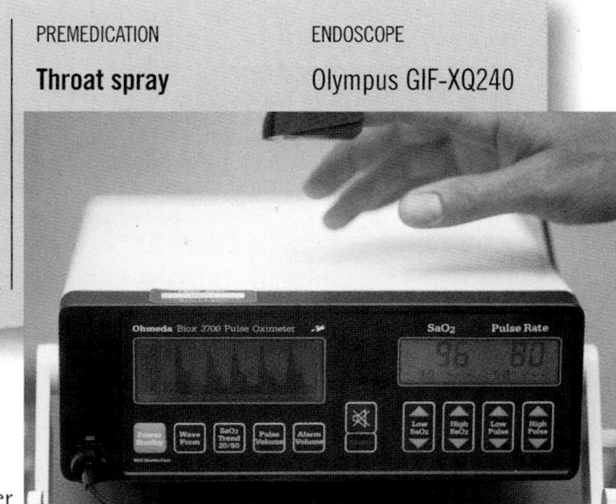

Pulse oximeter

36.1 Complete the table with words from A, B and C opposite.

Verb	Noun
consent	
	excision
incise	
	insertion
recover	
swallow	

36.2 Find words in A and B opposite with the following meanings.

1 to pass (an instrument through an orifice)
2 a substance used in procedures for lubrication
3 the flexible part of the endoscope
4 to stop something bleeding by applying heat
5 a growth that protrudes from a mucous membrane
6 to remove diseased tissue
7 taking a sample of a tissue for analysis
8 not flexible
9 a drug that numbs a particular part of the body
10 become accustomed to

36.3 Replace the underlined words and phrases with alternative words and phrases from C opposite.

After connecting the patient to an (1) <u>instrument which measures levels of oxygen in the blood and pulse rate</u> and placing him (2) <u>on his left side</u>, oxygen was provided through a (3) <u>tube in his nose</u> and the (4) <u>drug treatment prior to the procedure</u> administered as stated. Shortly afterward, the endoscope was (5) <u>inserted</u> into the oesophagus. After the examination, the patient was (6) <u>moved</u> to the recovery area.

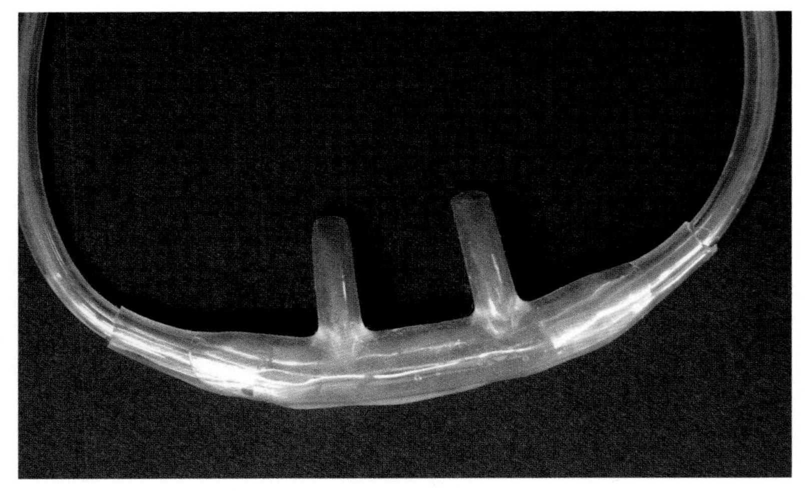

Nasal cannula

Over to you

What would you say to a patient during a bronchoscopy, or during another internal investigation which you carry out regularly?

37 X-ray and CT

A Radiography and radiology

Radiography involves exposing a part of the body to a small dose of **radiation** to produce an image of the internal organs. Organs with high density such as the ribs and spine are **radiopaque**, meaning they absorb radiation, and appear white or light grey on the image. Lung tissue is **radiolucent** – doesn't absorb radiation – and appears dark on the image. Before some types of **X-rays**, patients are given a liquid called a **contrast medium**, such as barium or iodine, which is radiopaque and allows the organ or tissue it fills to be examined. The contrast medium may be swallowed, introduced through the anus as an **enema**, or given as an injection.

Radiology is the use of radiation in the diagnosis and treatment of diseases such as cancer.

B X-ray examination

The chest X-ray is the commonest diagnostic X-ray examination. Normally a frontal (anteroposterior) view is obtained. The patient stands **facing** the photographic plate with the chest pressed to the plate, with hands on hips and elbows **pushed out** in front. The **radiographer**, the technician who takes the X-ray, asks the patient not to move, then to breathe in deeply and not to breathe out. This makes a **blurred**, unclear X-ray image less likely and improves the quality of the image, as it is easier to see **abnormalities** in air-filled (**inflated**) lungs than in **deflated** lungs.

Keep still.

Now, **take a deep breath** and **hold your breath.**

For a side, or lateral view, the patient is asked to **stand sideways** to the photographic plate with **arms raised.** A chest X-ray may be repeated at intervals to track for any changes. These repeated examinations are called **serial** chest X-rays.

C Computed Tomography

Here is an extract from a hospital's press release.

The Western General has installed a new GE LightSpeed 16 Computed Tomography **(CT) Scanner**. CT uses an X-ray source which rotates around the body to produce cross-sectional images.

The new scanner takes up to 16 simultaneous cross-sectional images (**slices**) and transmits more data in less time than ever before. Each slice can be less than one millimetre thick, making it possible to find very small abnormalities.

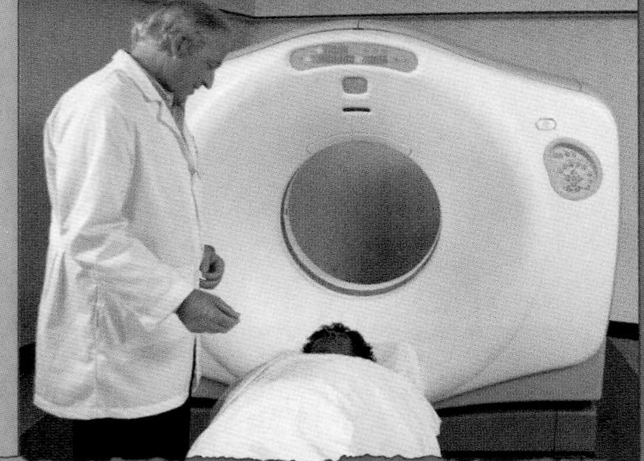

The scanner will be used for:

- diagnosing muscle and bone disorders
- locating tumours, infections and clots
- monitoring the progress of malignant diseases and the **response to therapy** (treatment)
- providing accurate guidance for **interventional procedures** such as biopsies, and **drainage** – removing fluid from the site of an injury or infection.

37.1 Choose the correct word or phrase to complete each sentence. Look at A and B opposite to help you.

1 In radiography, barium is used as a contrast .... (image/medium).
2 Tissue which absorbs radiation and appears white on an X-ray is .... (radiolucent/radiopaque).
3 An .... (enema/injection) is a liquid introduced into the bowel by way of the anus.
4 X-rays used to measure the progress of a disease are called .... (repeated/serial) X-rays.
5 If a patient moves during an X-ray, the image may be .... (blurred/abnormal).
6 It's easier to see abnormalities when the lungs are .... (deflated/inflated).

37.2 Complete the words. Each begins with *radio*. Look at A and B opposite to help you.

1 Using radiation to diagnose and treat disease is radio.............................. .
2 Using radiation to make images is radio.............................. .
3 Using radiation to provide treatment is radio.............................. .
4 If something is not penetrable by radiation, it is radio.............................. .
5 If something is penetrable by radiation, it is radio.............................. .
6 A technician who administers X-rays is a radio.............................. .

37.3 Complete the radiographer's instructions. Look at B opposite to help you.

Please stand (1) .... this board.
Put your hands on the back of your hips and your elbows forward. I'll help you.
(2) .... your elbows (3) ....
Keep (4) ....
In a moment I'll ask you to (5) .... a deep breath in and hold it.
Breathe in, (6) .... it.
That's it. Fine. You can breathe out now.
Thank you. I'll need to check the film.
Now I'm going to take a side view.
Can you stand (7) .... with your right side close to the machine and your arms raised?

37.4 Complete the table with words from B and C opposite.

Verb	Noun	Adjective
		abnormal
breathe		
drain		
intervene	intervention	
		therapeutic

Over to you

Explain to a patient what an X-ray is and what a CT scanner does.

38 MRI and ultrasound

A Magnetic Resonance Imaging (MRI)

MRI is safer than X-rays because it does not use **radiation.** It provides more information than Computed Tomography (see Unit 37) about some head, neck and spinal disorders because the images are more detailed and have more **contrast**, meaning the differences between dark and light areas are stronger. Unlike CT, the images can be taken on any **plane** – any surface of the body seen from an angle. It is now used for **imaging** – taking images of – the brain and heart, and in oncology.

Contraindications, cases in which it should not be used, include patients with metallic **foreign bodies** in the orbits, and patients with **pacemakers** – electronic devices surgically implanted to regulate heartbeat. MRI is also not approved during the first **trimester** of pregnancy.

B Ultrasound

Ultrasound examination uses **high-frequency sound waves** to view organs and structures inside the body. The waves are generated and received by a hand-held device called a **transducer.** The reflected waves are processed by a computer which produces detailed images for display on a monitor. Ultrasound is safe as it does not employ ionizing radiation like X-rays. It is a cheap, quick and **non-invasive** investigation – with no surgical procedure – for a wide range of **referrals**, although results can be unsatisfactory in **obese** (overweight) patients.

C Preparing for medical imaging

Preparing for an MRI scan

You will need to have **completed a safety questionnaire** and have it with you(1). It is important that there is no metal on your clothing or person(2). Prior to the scan you can eat and drink normally(3). Your details and safety questionnaire will be checked with you by the radiographer, who will explain the procedure and answer any questions you may have(4). You will be asked to remove any **metallic objects**, as well as credit cards(5).

You will be asked to lie on the MRI **scanner table** and **make yourself comfortable**(6). The radiographer will **position** the part to be scanned carefully in the scanner(7). During the MRI scan you will not feel anything but you will be required to **stay still** to achieve the best possible images(8). The whole examination process takes approximately 45 minutes(9). Depending on the site, you may require an injection (10). Only the part of your body requested by your doctor or source of referral will be scanned.

Preparing for an ultrasound

You will be asked to lie on an examination table(11). A special **gel** is **applied** to your skin(12). This ensures there are no air pockets between the transducer and your body(13). The transducer is moved over the area to be examined(14). You may feel some pressure and **experience some discomfort**, especially if the test requires you to have a full bladder(15). You may be asked to change your position for clearer pictures(16). When the radiologist is satisfied with the picture quality, the test is done and the gel is **wiped off**(17). A typical test may take between 20 minutes and one hour(18).

D Describing medical imaging

An ultrasound scan of the liver **revealed** reduction of metastases.

An ultrasound scan of the abdomen **demonstrated** a small right renal tumour.

An ultrasound scan **showed** an intra-abdominal abscess.

38.1 Match the two parts of the sentences. Look at A opposite to help you.

1 MRI provides more detailed information than CT because
2 MRI is not approved for use in
3 MRI is safer than X-rays because
4 MRI allows imaging on many planes

a there is no radiation.
b unlike CT.
c of high contrast sensitivity.
d the first three months of pregnancy.

38.2 Match what the radiographer says during an MRI scan with a numbered point in C opposite.

a You might need an injection.
b I want you to lie down and just relax.
c It's important that you try not to move.
d I'm going to go through your questionnaire with you.
e It will be over in three-quarters of an hour.
f It's very important that you put any metal objects into this tray.

38.3 Match what the radiographer says during an ultrasound with a numbered point in C opposite.

a I'm going to put some gel on your abdomen. You might find it a bit cold.
b That's it. All done. I'll just clean you up.
c I'd like you to lie flat on your back on the table.
d The gel is to make sure there's a good contact with your skin.
e I'll move this back and forwards to cover the whole area.

38.4 Make word combinations using a word from each box. You may need to look at Units 34 to 37. Then use some of the word combinations to complete the sentences.

breathe	anaesthetic
excise	your breath
experience	area
foreign	diseased tissue
hold	discomfort
informed	in
introduce	bodies
local	consent
recovery	the endoscope

1 I'm going to give you a .... so that you won't feel any pain.
2 With an MRI, it's important there are no metallic .... in the eyes.
3 After an operation, patients are moved to a .... to recuperate.
4 Endoscopes can be used to ....
5 Before an endoscopy, the patient's .... must be obtained.

Over to you

Explain to a patient why you are referring her for an ultrasound scan or an MRI scan, and what she can expect to happen during the procedure.

39 ECG

A Uses of an ECG

An **electrocardiogram** (ECG) is a **tracing**, or drawing, produced by an **electrocardiograph** – a device which records electrical activity in the heart. An ECG can be used for:

- deciding if the heart is performing normally or suffering from abnormalities, for example cardiac arrhythmia – extra or **skipped heartbeats**
- indicating damage to heart muscle, such as heart attacks, or ischaemia of heart muscle (**angina**)
- detecting **conduction abnormalities**: heart blocks and bundle branch blocks (BBB)
- **screening for** ischaemic heart disease during an **exercise tolerance test**, often carried out on an exercise bike or treadmill
- providing information on the physical condition of the heart, for example in patients with left ventricular hypertrophy (LVH)
- detecting **electrolyte disturbances**, for example low plasma potassium levels.

B ECG procedure

Here is an extract from a medical textbook.

1. The patient should lie down and relax.
2. **Calibrate** the ECG machine – a standard signal of 1mV should move the **stylus** two large squares (1 cm) vertically.
3. Attach the **limb leads**: left arm **(LA)**, right arm **(RA)**, left leg **(LL)**, and right leg **(RL)**.
4. **Record** the six standard leads: **I**, **II**, **III**, augmented voltage right arm **(AVR)**, augmented voltage left arm **(AVL)**, and augmented voltage foot **(AVF)** – three or four **complexes** (see C below) for each.
5. Apply the electrode to the six **chest positions** in turn, recording three to four complexes of each. If the rhythm does not appear to be **sinus** (normal rhythm), a rhythm strip of 6–10 complexes in a single lead should be recorded.

C A normal ECG

The picture shows an ECG tracing of a normal heartbeat showing a P wave, a QRS complex and a T wave. Each large square is equivalent to 0.2 seconds. The R–R interval gives the **heart rate**, in this case 75/min. In the case of abnormalities, the QRS complex can be **widened** or too tall. The ST segment can be **elevated** or **depressed**. The T wave can be the right way up, or **inverted** – the wrong way up.

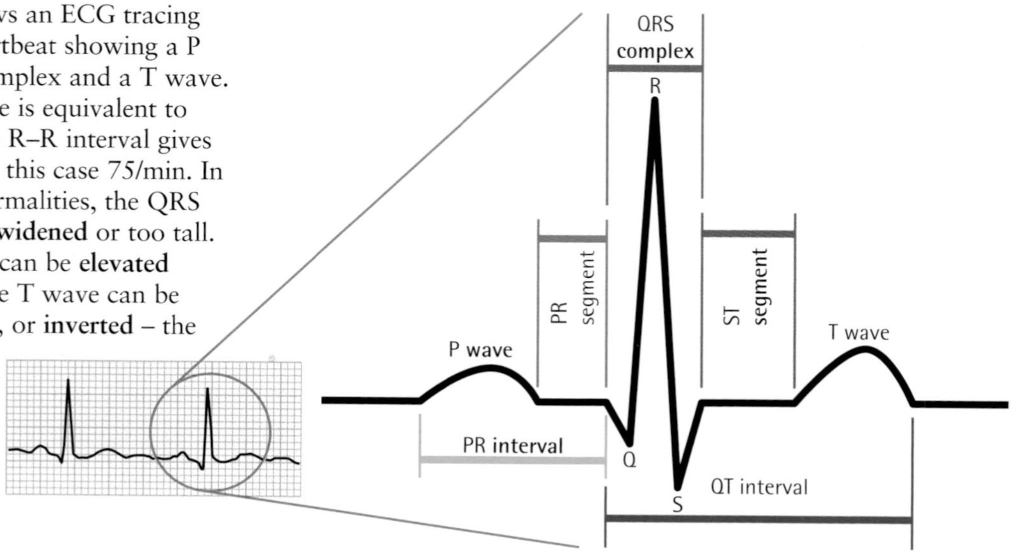

39.1 Find words and phrases in A and B opposite with the following meanings.

1 the marks produced by an ECG stylus
2 a test which determines how well a patient copes with physical exercise
3 a missed heart beat
4 a change in the chemical composition of body fluids
5 the flow of electric current in the heart
6 testing for disease
7 check or adjust an instrument before use
8 the pen which produces the drawing

39.2 Label the limb leads (a–d) on the first diagram, and write a title (e) for the second diagram. Look at B opposite to help you.

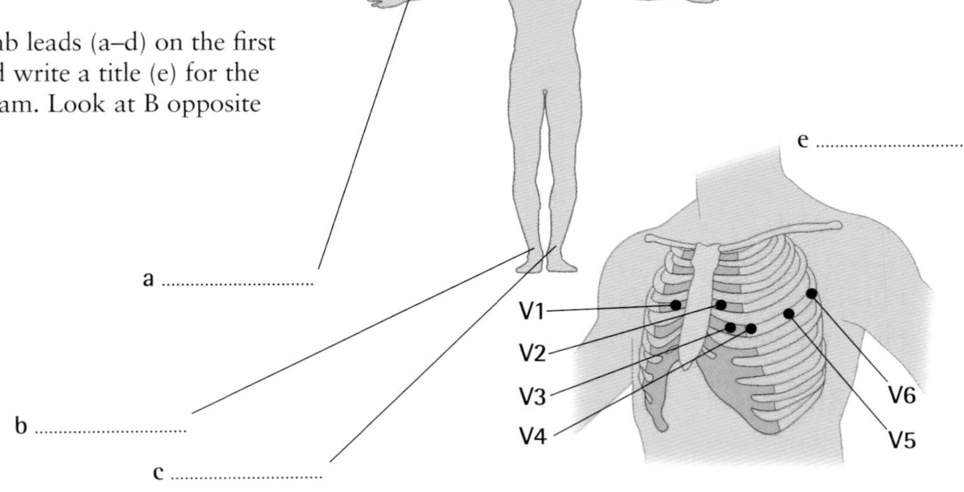

39.3 Complete the text using words from the box. Look at C opposite to help you.

complexes	leads	wave	interval	rate

This very abnormal ECG shows a (1) .... of approximately 33/min; a single long pause of approximately 4 seconds between ventricular complexes with atrial activity; widened QRS (2) .... in keeping with (R)BBB. Deep T (3) .... inversion in II, III, AVF and some chest (4) .... (V4–V6). Deep QRS complexes in V2 and V5 in keeping with LVH. One atrial ectopic. QT (5) .... is normal.

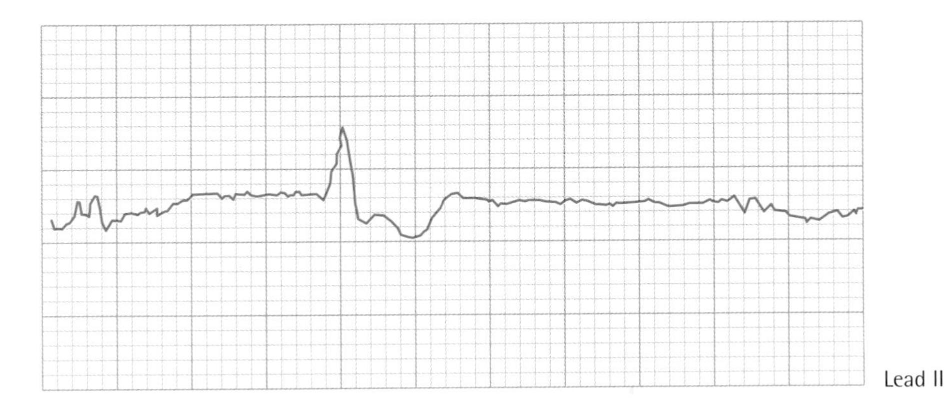

Over to you

Describe an ECG tracing of one of your patients to a colleague.

40 Medical treatment

A Prescriptions and drugs

As part of treatment, a doctor may **prescribe medication**, commonly referred to as **medicine** or **drugs**. A **prescription** may take these forms:

Tab. Nabumetone 500 mg
mitte 56
sig. 2 tab. Nocte

Old style

Nabumetone Tab 500 mg
send 56
label 2 tablets at night

New style

In the UK, patients take prescriptions to a **chemist's** shop, which sells a wide range of **non-prescription medicines** and other products such as cosmetics, for **dispensing** by a **pharmacist** (the person who prepares the medicines). In hospitals, prescriptions are dispensed by the **pharmacy** (the department where the drugs are prepared).

Drugs come in many different forms. See Appendix III on page 143 for descriptions and illustrations, and Appendix II on page 131 for a list of abbreviations used in prescriptions.

Note: Latin abbreviations in prescriptions are being replaced by English, but may still be seen.

B The British National Formulary

The British National Formulary provides information on prescribing and administering **prescription drugs** in the UK.

SULCONAZOLE NITRATE

Indications: Fungal skin infections

Cautions: Contact with eyes and mucous membranes should be avoided.

Side-effects: Occasional local **irritation** and hypersensitivity reactions include mild burning sensation, erythema, and itching. Also **blistering**. Treatment should be discontinued if these are severe.

Dose: Apply 1–2 times daily, continuing for 2–3 weeks after lesions have healed.

Exelderm® (Centrapharm)
Cream, sulconazole nitrate 1%

An indication is a situation or a sign that suggests a specific treatment should be given. A **contraindication** is a situation or sign that a specific drug or treatment should not be used or is **contraindicated**.

Exelderm® is a **proprietary** – commercial – name for a medication containing sulconazole nitrate. The same drug may have both a proprietary name and a **generic** name. For example, Prozac and Fluoxetine are the proprietary and generic names for the same drug.

40.1 Match the abbreviations (1–9) with their meanings (a–i). Look at Appendix II on page 131 to help you.

1 p.c.
2 q.d.s.
3 s.c.
4 s.l.
5 p.o.
6 c.c.
7 p.r.n.
8 i.v.
9 infus^{n}

a by mouth
b sublingual
c with meals
d as required
e after food
f intravenous
g four times a day
h infusion
i subcutaneous

40.2 Complete the sentences. Look at A and B opposite to help you.

1 .... of the skin may be caused by drugs such as aspirin which can produce a rash.
2 At a .... you can get your prescription and all sorts of other health products.
3 Gastro-intestinal irritation is a ....-.............................. of aspirin.
4 Aspirin is .... for patients with previous or active peptic ulceration.
5 When bubbles appear on the skin due to heat or irritation, this is called ....
6 The maximum .... of paracetamol for an adult is 4 grammes daily.
7 .... means a drug is not contraindicated but care must be taken in its use.
8 .... for codeine phosphate are mild to moderate pain and cough suppression.
9 A person who dispenses drugs is a ....
10 The place where drugs are dispensed in a hospital is a ....

40.3 Describe each of these prescriptions for a patient with suspected acute coronary syndrome. Look at Appendix II on page 131 to help you. The first one has been done for you.

Medicine	Dose	Method of administration
Streptokinase	1 500 000 U	i.v. infus^{n} over 60 mins
Aspirin	300 mg	p.o. stat
Diamorphine	2.5–5 mg	i.v. stat
Metoclopramide	10 mg	i.v. stat
GTN	300 mcg/5 ml	i.v. infus^{n} start @ 40 mcg/min

Streptokinase, one and a half million units by intravenous infusion over sixty minutes.
....
....
....

Practise writing prescriptions in English for medication you often have to prescribe in your own language.

41 Surgical treatment

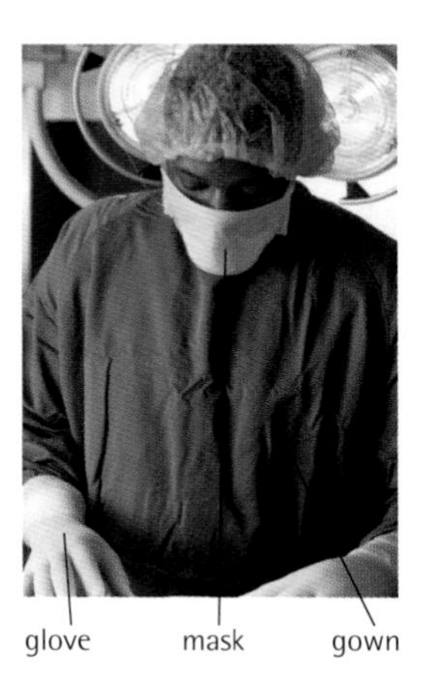

A The operating theatre

Surgery is carried out in an **operating theatre.** Great care is taken to make sure that operations take place in **sterile conditions** – free from microorganisms. The surgeon and his or her **assistant** wash or **scrub up**, and put on **surgical gowns, masks,** and **gloves.** The patient's skin is prepared by disinfecting it with an antiseptic solution. This is known as **prepping** (preparing) the patient. They are then covered with sterile **drapes,** so that only the area of the operation is exposed.

B Instruments

The most basic surgical instruments are shown in the picture.

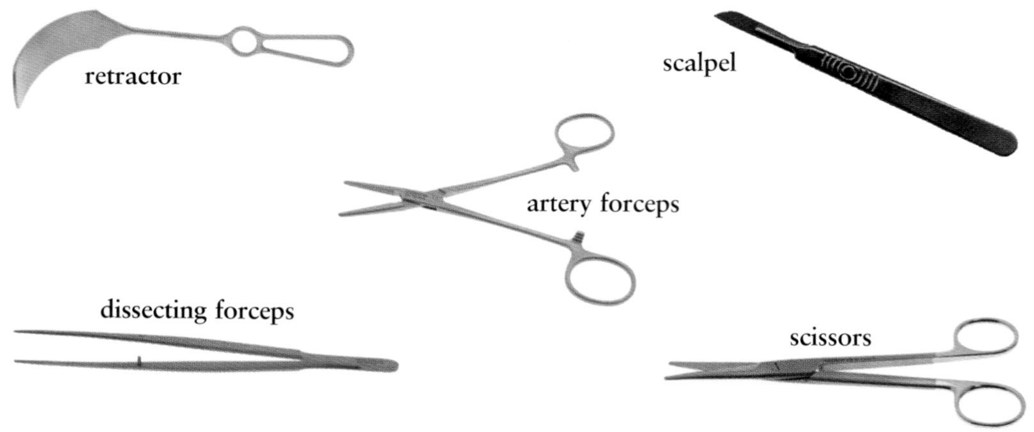

C The operation

The **operation** begins when the surgeon **makes an incision** or cut. Control of bleeding is very important. A **swab** is a pad of cotton or other material used to soak up blood from the operation site. A **sucker** is a mechanical device which **aspirates** – sucks up – blood. Bleeding vessels are tied with **ligatures** or sealed by an electric current (**diathermy**).

Drains may be inserted to carry away fluid which might act as a culture medium for bacteria. Various methods are used to close the wound, for example **sutures** (also known as **stitches**), or **staples.** Finally, the wound is **covered** with a **dressing.**

D An operation report

This patient had an indirect right inguinal hernia.

Anaesthesia: Spinal anaesthetic with local anaesthetic **infiltration**

Incision: Right inguinal

Procedure: The external oblique aponeurosis was **divided** and the spermatic cord **mobilized**. The hernial sac was identified and separated from the spermatic cord. The hernial sac was then mobilized back to its neck where it was **transfixed** and the **redundant tissue excised**. The fascia at the neck of the spermatic cord was divided, carefully **preserving** the vessels, the genital branch of the nerve and the vas deferens. Thereafter the posterior wall of the inguinal canal was **repaired** in two layers.

Closure: The wound was **closed in layers** with Dexon suture material to the external oblique aponeurosis and staples to the skin.

41.1 Which of the instruments shown in B opposite is needed for each of the following procedures?

1 making an incision
2 keeping the sides of the wound open
3 cutting sutures
4 holding the cut ends of blood vessels before they are tied

41.2 A surgeon is talking to a medical student about assisting at operations. Complete his advice using words from A, B and C opposite.

An (1) .... must be able to carry out the following tasks to help the surgeon. Firstly, he or she must help in (2) .... the patient and putting the (3) .... in place to provide (4) .... conditions. Expert handling of a (5) .... is essential to allow the surgeon to see what he is doing. The assistant must also keep the operation site free of blood, by careful use of the (6) .... or (7) .... The surgeon also needs assistance with tying and cutting (8) .... , and with the insertion of a (9) .... , if necessary. Finally, the assistant may be asked to close the wound with (10) .... or other devices.

41.3 Find words in C and D opposite with the following meanings.

1 cut into two parts
2 corrected (something that was damaged)
3 freed from surrounding tissues
4 removed by cutting out
5 spread of liquid into an area
6 making sure something is not damaged
7 small metal devices to hold the edges of a wound together
8 unnecessary
9 sewing up of the wound
10 flat, thin pieces of tissue that lie on top of one another

Over to you

Keyhole surgery (endoscopic surgery) has become popular. How important is it in surgery? Why?

42 Therapies

A Radiotherapy and chemotherapy

Radiotherapy is the use of radiation in controlled doses to treat cancers. It works by damaging the DNA of malignant cells. Radiotherapy may be used:

- as **curative** treatment, for example to shrink tumours
- as **adjuvant** treatment, alongside or following **chemotherapy** – treatment with anti-cancer drugs
- in lower doses as **palliative** treatment to reduce pain and other symptoms of cancer or disease, but not as a cure.

A **radiologist** determines the dose and the exact target for the radiation beams. Dosage is measured in **grays** (**Gy**). A daily dose is a **fraction**. Radiotherapy can also be delivered internally by radioactive **implants** such as needles, or by liquids such as strontium for some bone cancers.

B A day in the life of a physiotherapist

Sam is a hospital **physiotherapist**. She works mainly with patients who have conditions or injuries affecting the lower extremities such as fractures, torn ligaments, and cartilage tears. Most of her patients are **referrals** from other departments in her hospital. She also works with patients **in rehabilitation** following orthopaedic surgery. Some are young people with sports injuries, others are elderly people who have had **hip replacements.** Among her therapies are **manipulation, massage,** and exercise to keep the joints mobile and to strengthen muscles. **Rehabilitating** some patients means helping them to walk again using **crutches** or **Zimmer frames.**

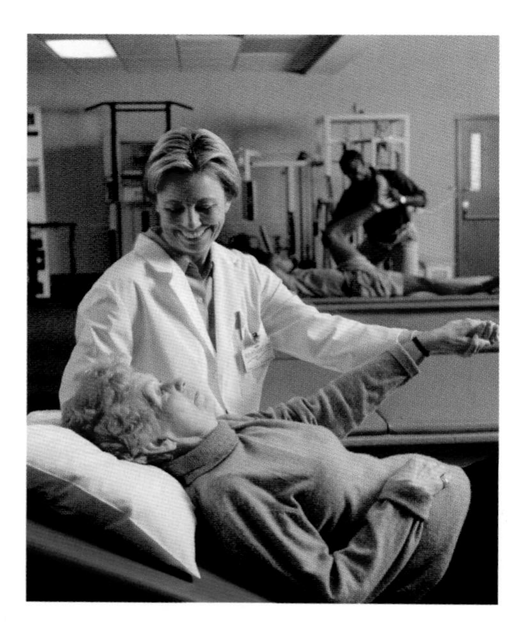

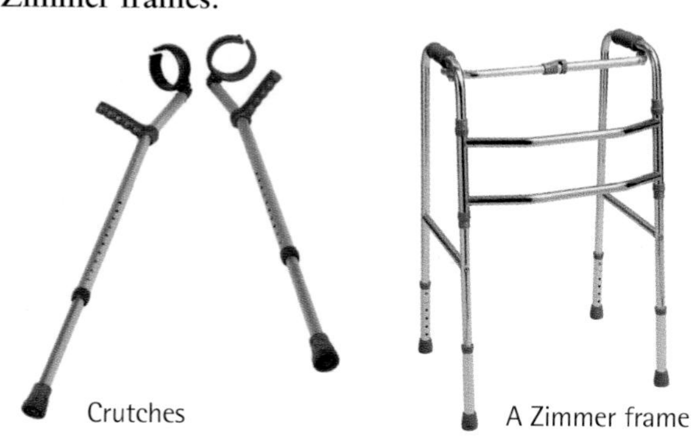

Crutches

A Zimmer frame

C Cognitive Behavioural Therapy

Cognitive Behavioural Therapy (**CBT**) is one of the 'talking therapies' for mental health conditions. It aims to eliminate negative thoughts and change behaviour in response to these thoughts. It can help with anxiety, panic attacks, depressive disorders, **post-traumatic stress disorder** and **chronic fatigue.** Therapy can be provided in **face-to-face sessions** with a **therapist**, but delivery by computer can also be effective. It is more effective than **counselling** for some and can provide long-term protection against **relapse**, a return of symptoms after improvement. However, patients must be committed to solving their problems and prepared to work on them between sessions.

42.1 Name the therapy being described. The same therapy may be described more than once. Look at A, B and C opposite to help you.

1 Treatment with drugs toxic to cancer
2 Treatment of cancer by radiation
3 May include massage
4 Aims to eliminate negative thoughts
5 Can involve helping people to walk again
6 Can help with panic attacks

42.2 Complete the sentences. Look at A opposite to help you.

1 .... treatment is given in larger doses than palliative treatment.
2 Radiotherapy can be used alongside other treatments as .... therapy.
3 .... treatment is treatment which helps relieve the symptoms of a condition but does not cure it.
4 Radioactive .... are wires or needles placed into the area to be treated.
5 A typical .... for an adult is 1.8 to 2 Gy.

42.3 Complete the text describing a typical working day for Sam. Look at B opposite to help you.

Work begins around 8.00 am. I check for new (1) .... on the computer and prepare for my morning appointments. Between 9.00 and 12.00 I see new patients around the hospital. I carry out assessments on them and decide what the appropriate form of (2) .... is. I work with broken limbs, joint (3) .... and ligament repairs. I deal with A&E patients as well as patients referred to me by their GPs or specialists.

After lunch I see regular patients. Their therapy includes exercises to increase range of motion and to strengthen muscles.

What do I like about being a (4) .... ? Getting people back to normal. (5) .... people so they can get on with their lives after an injury or an operation.

Over to you

Talk about two contrasting therapies you have experienced, and the advantages and disadvantages of each.

43 Screening and immunization

A Screening

Screening is a way of identifying people at **increased** or **greater risk** for a condition, although they do not yet have any signs or symptoms. In some cases, **mass screening** – screening large numbers of people – is appropriate, for example in the past for tuberculosis. In other cases, only those with **high risk factors**, like a **family history** of conditions such as cancer and diabetes, are screened. However, there are a number of problems with screening. There are always **false negatives**, cases where a patient has a disease but screening does not identify it. There are also **false positives**, where someone is told they have a disease when in fact they do not. Furthermore, with some diseases, early identification is of no benefit to the patient as there is no treatment available.

B Common screening tests

Condition	Test	Subjects	Frequency
Neural tube defects and Down's Syndrome risk	**AFP**	pregnant women	between 16 and 17 weeks
Breast cancer	**mammography**	women, 50–70	every 3 years
Cervical cancer	**smear test**	women, 20–60	every 3 years
Cardiovascular disease	**blood cholesterol**	>40 with high risk factors	every year
Secondary prevention			
Cholesterol >4	**blood cholesterol**	patients with heart disease	every 6 months
Diabetic retinopathy	**ophthalmoscopy** (see Unit 34)	patients with diabetes	every year

C Immunization for travellers

The following **vaccinations** are recommended for travellers to South Asia:

HEPATITIS A OR IMMUNE GLOBULIN (IG)
Transmission of hepatitis A virus can occur through direct **person-to-person contact**; through **exposure to** contaminated water, ice, or shellfish harvested in **contaminated water**; or from uncooked fruits, vegetables, or other foods.

HEPATITIS B
Especially if you might be **exposed to** blood or **body fluids** (for example, health-care workers), have sexual contact with the local population, or be exposed through medical treatment.

JAPANESE ENCEPHALITIS
If you plan to visit rural farming areas and under special circumstances, such as a known **outbreak** of Japanese encephalitis.

MALARIA
Your risk of malaria may be high in these countries, including cities. Travellers should take an effective antimalarial drug.

RABIES
If you might have extensive, **unprotected** outdoor exposure in rural areas.

TYPHOID
Typhoid fever can be **contracted** through contaminated drinking water or food. Large outbreaks are most often related to **faecal contamination** of water supplies or foods sold by street vendors. Vaccination is particularly important because of the presence of S. typhi strains **resistant to** multiple antibiotics in this region.

AS NEEDED
Booster doses for tetanus, diphtheria and measles, and a one-time dose of polio for adults.

43.1 Complete the sentences. Look at A, B and C opposite to help you.

1 .... for heart disease include smoking, high cholesterol and a family history of heart disease.
2 In a small number of cases screening will not identify patients with the early signs of a disease. These are ....
3 Some people without signs of the disease will be wrongly identified as having the disease. These are ....
4 People .... blood or body fluids should be immunized against Hepatitis B.
5 A .... vaccination is given some time after the first vaccination to make sure the level of antibodies remains high.
6 Hepatitis B can be .... through exposure to body fluids.
7 Penicillin now has no effect against some hospital-acquired infections as they are .... penicillin.
8 An .... of measles has affected a number of children who had not been immunized with the MMR vaccine.

43.2 Complete the sentences using information from B opposite. The first one has been done for you.

1 Women aged from 50 to 70 *should have mammography every three years to check for breast cancer* .
2 Patients with heart disease ....
3 Women between 20 and 60 ....
4 Patients over 40 with high risk factors ....
5 Patients with diabetes ....
6 Pregnant women ....

43.3 Which of the immunizations listed in C opposite would you recommend for the following visitors to South Asia?

1 A tourist who will stay for a few nights in five-star hotels in major cities
2 A backpacker who will travel by local transport from one city to another
3 A medical student doing an elective in a city hospital
4 A volunteer who will live for a year in a rural community
5 A traveller who has not had a tetanus vaccination for ten years

Over to you

What immunizations are advised for visitors to your country, or a country you would like to visit?

44 Epidemiology

A Rates

The study of the spread and control of diseases in the community (**epidemiology**) requires analysis of **frequency** – the number of times something occurs in a particular period. We talk about frequency using word combinations with **rate**:

birth	**rate**	= the number of births in a **population** (group of people)
death (mortality)	**rate**	= the number of deaths in a population
case fatality	**rate**	= the number of deaths from a particular disease
survival	**rate**	= the number of patients still alive after treatment for a particular disease

The birth rate in Singapore is 10.2 per 1,000 population.

The infant mortality rate in Singapore is 3.0 per 1,000 live births.

Other words used to talk about frequency:

Malaria is	**common**	**in** many developing countries.
	uncommon **rare**	**in** developed countries.

B Incidence and prevalence

The **prevalence** of a disease is the number of people in a population who have the disease at a particular time. The **incidence** of the disease is the number of new cases of the disease during a particular time.

The incidence of malaria is	**high**	in many developing countries.
	low	in developed countries.

The incidence of HIV infection is **highest** in sub-Saharan Africa.

The incidence of HIV infection is **lowest** in the Caribbean.

Note: You don't say 'The incidence of malaria is common.'

The incidences of a disease in different groups are often compared.

Rheumatoid arthritis is **more common** in females **than** in males.

Rheumatoid arthritis is **less common** in males **than** in females.

Men and women are **equally affected by** restless leg syndrome.

C Association and causation

Studying the incidence of disease in particular groups of people can **lead to** the discovery of **causation** – what makes it occur. For example, the high incidence of lung cancer among cigarette smokers suggests that smoking is a **causative factor** in the development of lung cancer. However, care must be taken not to assume every **association between** a disease and a measured variable is **causative**. To say there is an association between two things simply means that they occur together in a certain situation. For example, several studies have indicated high rates of lung cancer in cooks. Although this could be a consequence of their work (perhaps caused by carcinogens in fumes from frying), it may be simply because professional cooks smoke more than the average. In other words, smoking might **confound** the association with cooking.

44.1 Complete the sentences. Look at A and B opposite to help you.

1 The proportion of episodes of illness that end fatally is the .... rate.
2 The proportion of people who die in a particular period is the .... rate.
3 The proportion of babies born in a particular period is the .... rate.
4 The proportion of patients who are still alive five years after treatment is the five-year .... rate.
5 Ten per cent of the population have perennial rhinitis at any time. This is the ....
6 Every year, 0.5–1.0% develop the condition. This is the ....

44.2 Complete the text, using the correct grammatical form of each word. Look at the table and at A, B and C opposite to help you.

Incidence of lung cancer
(Surveillance and Risk Assessment Division, CCDP, Health Canada)

Country	Incidence (per 100, 000)	
	Men	Women
China	67.5	26.6
Canada	59.1	30.8
USA	55.7	33.5
UK	51.2	22.0
Japan	44.6	13.3
Sweden	22.0	12.9
India	12.1	3.8

In all countries, lung cancer was (1) .... in men than in women. The (2) .... incidence among men was in China where 67.5 men per 100,000 were diagnosed with lung cancer in the period. In women, however, lung cancer was less (3) .... in China than in the USA and Canada.
The (4) .... incidence in both men and women was in India. Sweden also had a fairly (5) .... incidence in both sexes. Lung cancer is probably (6) .... in India for cultural and economic reasons, whereas in Sweden it is the result of an effective health education campaign.

44.3 Find prepositions in A, B and C opposite that can be used to make word combinations with the words in the box. Then use the correct forms of the words to complete the sentences.

affected	association	incidence	lead	rare

1 Recent surveys showed that the .... HIV cases had come down in the southern states of India.
2 Epidemiologic data suggest an .... obesity and depression, but further studies are needed.
3 Appropriate preventive measures should .... a reduction in the mortality rate.
4 Thalassaemia is .... people of Northern European origin.
5 Male and female soldiers are equally .... post-traumatic stress disorder.

Over to you

Which diseases have the highest incidence in your country? What is the commonest cause of death in men and in women?

45 Medical ethics

A GMC guidelines

care = protecting and looking after someone

professional competence = level of specialized knowledge and skills

confidential = private, secret

prejudice = have a negative influence on

risk = possibility of something bad happening

fit to practise = in a suitable condition to work

The duties of a doctor registered with the General Medical Council include:

1. Make the **care** of your patient your first concern.
2. Treat every patient politely and considerately.
3. Give patients information in a way they can understand.
4. Keep your professional knowledge and skills up to date.
5. Recognize the limits of your **professional competence**.
6. Be honest and trustworthy.
7. Respect and protect **confidential** information.
8. Make sure that your personal beliefs do not **prejudice** your patients' care.
9. Act quickly to protect patients from **risk** if you have good reason to believe that you or your colleague may not be **fit to practise**.

Note: For a full list of the guidelines, see www.gmc-uk.org

B Bioethical issues

Euthanasia
Should the medical profession help the terminally ill to **end their lives** when they choose?

Genetic engineering
Should we permit an **embryo** to be **cloned** – copied exactly – to replace a child who has died? Should parents be able to select the **genetic makeup** of their children to produce so-called **designer babies**?

Human fertility
IVF – in vitro fertilization – has made it possible for infertile women to have children, but should this include women long past the normal age of childbearing?
Embryos can be **frozen** and implanted in the mother at a later date but should this require the **consent** or permission of both parents if the marriage has broken down?
What are the rights of a **surrogate mother**, one who carries a child for a woman who is unable to do so, over that child?

Transplant surgery
Who should **give consent** for the removal of body parts for transplant surgery?

C Assisted dying

A medical student has made some language notes on a journal article.

A 53-year old woman with **incurable** muscular dystrophy flew to Switzerland to end her life. **Assisted dying** is legal in Switzerland but illegal in the UK. Opponents of euthanasia, or 'mercy killing', argue that legalization would lead to abuse and call for doctors who participate to be **struck off**. What Britain needs, they claim, is better **palliative care** and more **hospices** for the **terminally ill** to allow such patients to die with dignity.

This follows a recent US case where the husband of a woman who had been in a **persistent vegetative state** for 16 years was successful in having artificial feeding withdrawn in spite of opposition from his wife's parents.

struck off = removed from the GMC register and banned from practising medicine in the UK

palliative care = treatment to relieve, rather than cure symptoms

hospice = facility providing care for terminally ill patients

terminally ill = not expected to live

persistent vegetative state = unable to speak or follow simple commands; does not respond in any psychologically meaningful way

45.1 Which of the GMC guidelines in A opposite is breached in each of these cases?

a A GP falls asleep regularly during consultations. His colleagues do nothing.
b A doctor is aware that a patient has a history of violence against women. She informs a friend whose daughter has just become engaged to this man.
c A doctor attempts to dissuade a patient from having an abortion as this procedure is against his religious beliefs.
d A doctor refers a patient to a medical textbook for an explanation of his pancreatic cancer.
e A doctor fails to complete the number of days of professional development training advised annually.
f A doctor tells a seriously overweight patient who has ignored his advice to diet that she deserves any ill effects that might result from her obesity.

45.2 Match each headline to an opening line from a newspaper report. Look at B and C opposite to help you.

1 **66-year-old becomes oldest mother**

2 **Frozen embryo case to go to Europe**

3 **Surrogate mother sued by couple**

4 **Embryo cloning – where will it take us?**

5 **Using body parts without consent**

6 **UK full face transplant search on**

7 **'Designer baby' rules are relaxed**

8 **Doctors back infant mercy killing**

a A woman of 30 who agreed to bear a child for a childless couple then refused to part with the child has been …
b A surgeon has been accused of removing organs from patients without their knowledge …
c Three-quarters of Belgian doctors are willing to assist in the death of critically ill babies to end their suffering.
d A woman hoping to stop the destruction of six embryos created with her eggs and her ex-partner's sperm launched a case at the European Court of Human Rights …
e A woman has given birth in Romania following IVF treatment.
f Consultant plastic surgeon Dr Peter Butler has been given the go-ahead by a hospital ethics committee to find a patient who meets selection criteria for a full face transplant.
g If your favourite pet dies, it is technically possible to produce exact replicas – but what about humans?
h It is now legal to select embryos to provide blood cell transplants for sick siblings.

Over to you

What are your views on assisted dying?

46 Research studies

A Case-control studies

Here are some extracts from a medical textbook.

In a **case-control study**, a group of people suffering from a disease (the **cases**) is compared with a group who do *not* have the disease, but are similar in other ways (the controls). The two groups, the participants in the study, are compared to see if they were **exposed** to a possible **risk factor** or not. A **risk factor** is something which may contribute to the cause of disease. This type of study is often used as a first step in identifying the cause of a disease.

B Cohort studies

In a **cohort study**, a group (**cohort**) of people (**subjects**) who are similar is studied over a period of years (a **longitudinal study**) to determine if there is a relationship between **exposure** to a risk factor and development of a disease. At the beginning of a **prospective** study, none of the subjects has the disease. They are **followed up** for a number of years, and at the end of the period, those who have developed the disease are compared with those who have not. In a **retrospective** study, the researchers look back, by studying hospital records for example, at what has happened in the past, comparing subjects who have developed the disease with those who have not. In a cohort study and in a case-control study the subjects are only observed and there is no **intervention** such as drug treatment or surgery. A cohort study is a more reliable method of identifying the cause of disease than a case-control study. But for proof of cause, a trial is needed.

C Trials

In a **trial**, a group of people who are suffering from a disease are given a particular treatment. To determine the effectiveness of the treatment, a **controlled trial** is performed. Two groups are studied: one group (the **study group**) is given the treatment and the other (the control group) is not. The controls may be given a **placebo** – something which seems to be identical to the treatment but which has no effect. If there is an equal possibility that patients may be selected for the study group or for the controls, the trial is said to be **randomized**. A **randomized controlled trial** makes **bias** – error in a study which influences the results – less likely. An additional way of removing bias is **blinding**: patients do not know if they are receiving the treatment or the placebo. If, in addition, the researchers do not know who is receiving the treatment, the trial is a **double blind trial**. Randomized controlled trials are used to test treatments or preventive measures.

D Variables

If the subjects in a study are all aged 50, then age is a **constant** in the study. If their ages range from 20 to 70, then age is a **variable**. A **confounding variable** is any variable which is associated with both the disease and the risk factor being studied (for example, smoking in the case of cooks and lung cancer discussed in Unit 44). If such variables exist there is no way for the researcher to know whether the difference in the risk factor or the confounding variable is the one that is truly causing the disease.

46.1 Complete the table with words from A, B and C opposite.

Noun	Verb
	bias
	control
exposure	
(person)	participate
	intervene
	study

46.2 Complete the sentences with a word from A, B, C or D opposite.

1 People who are not receiving the experimental treatment, but who are otherwise the same as those receiving it are ....
2 A trial in which neither the subjects nor the researchers know who is receiving the treatment is a ....-.. trial.
3 A study that follows the participants over many years is a .... study.
4 A .... is a group of people with similar characteristics.
5 Allocation to groups is .... if all participants have equal chance of being in either group.
6 A harmless substance given to some participants to test the effect of a trial substance is a ....
7 Something that might be a part of the cause of a disease is a .... factor.
8 Something that might cause confusion about the cause of a disease is a .... variable.

46.3 Study the research questions below, and in each case decide which of the research study types mentioned in A, B and C opposite would answer the questions best. Use the index or a dictionary to look up any unfamiliar words.

1 To examine the outcomes of an unwanted first pregnancy (abortion v live delivery) and risk of depression.

(*BMJ* 2005;331: 1303 Reproduced with permission from the BMJ Publishing Group)

2 To test the hypothesis that supplemental oxygen reduces infection risk in patients following colorectal surgery.

(*JAMA* 2005; 294: 2035)

3 To develop a relatively simple, inexpensive, and accurate test that measures telomerase activity in voided urine to apply to large-scale screening programs for bladder cancer detection.

(*JAMA* 2005; 294: 2052)

4 To evaluate the relative risk of being responsible for a fatal crash while driving under the influence of cannabis.

(*BMJ* 2005;331: 1371 Reproduced with permission from the BMJ Publishing Group)

Over to you

Describe a research study that you've carried out or would be interested in carrying out. In your experience, what are the main problems encountered in designing a research study?

47 Taking a history 1

A A full case history

A full case history covers:

- **personal details**
- **presenting complaint**
- **past medical history (PMH)**
- **drug history** (see Unit 48)
- **family history** (see Unit 48)
- **social and personal history** (see Unit 48)
- **patient ideas, concerns and expectations** (see Unit 49)
- **review of systems** (see Unit 49).

B Personal details

Normally, patients' personal details have been entered in their records by a nurse or administrative staff before a doctor sees them. However, on later consultations a doctor may wish to check details such as address, date of birth, occupation or marital status.

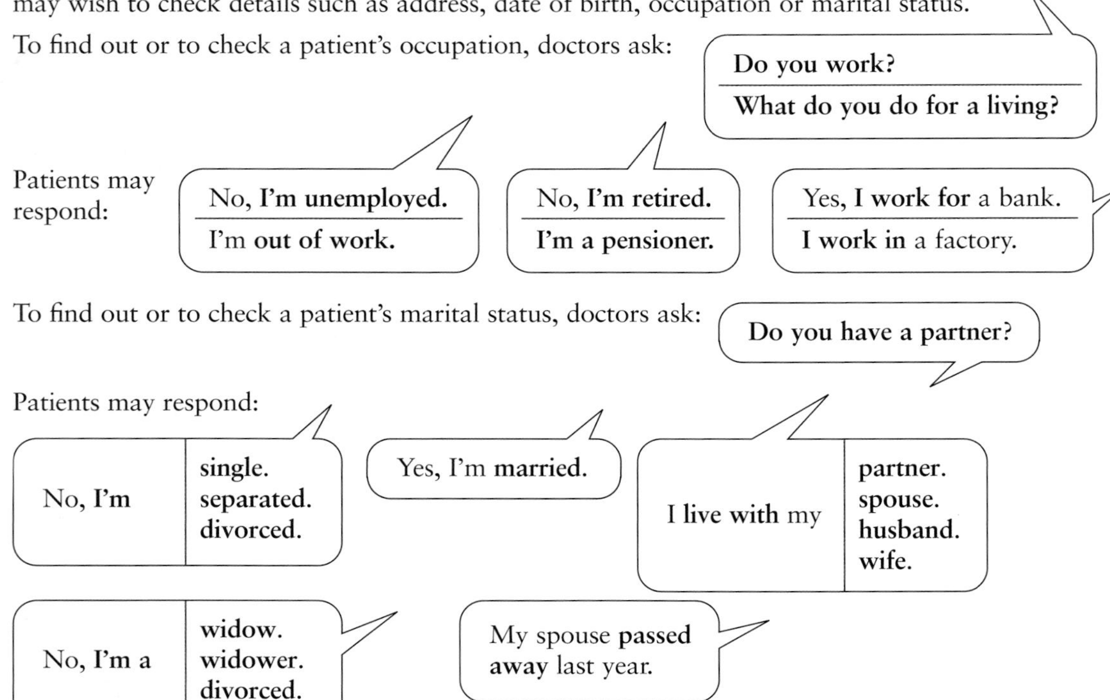

Note: A **spouse** can be a wife or husband. A **widow** is a woman whose husband is no longer living; a **widower** is a man whose wife is no longer living. To avoid saying someone is dead, patients may say that person has **passed away**.

C Talking about pain

Case notes are kept for each consultation. The Presenting Complaint section of case notes records the patient's symptoms. For example:

> R. sided temporal headache,
> severe, throbbing.
> Lasts 24-48 hrs.

In this case, the patient complains of a headache. For a list of the questions the doctor asked this patient, and for patients' descriptions of other kinds of pain, see Appendix IV on page 144.

47.1 Study the case notes. Write the questions the doctor asked to obtain the numbered information. Look at Appendix IV on page 144 to help you.

SURNAME *Oates*		FIRST NAMES *Allison*
ADDRESS *Acredales, Bridgethorpe*		
D.O.B. *30/4/79*	SEX *F*	MARITAL STATUS *married*(1)
OCCUPATION *bank clerk*(2)		
Presenting complaint *c/o severe headache, boring in nature(3), mainly in and around R eye(4).* *Can radiate to forehead(5). Comes on at any time(6) and can vary in duration 1–2hrs(7). No precipitating(8) or relieving(9) factors.* *Has noticed haloes around lights with some blurry vision in R eye and vomiting(10).*		

47.2 Read the continuation of the case notes for the patient in C opposite. Write the doctor's questions. Look at Appendix IV on page 144 to help you.

> *PMH similar headaches 10 yrs, coming every 3 mths. Often premenstrual. Aggravated by eating chocolate; relieved by lying in dark room. Can have visual aura, blurred vision, nausea + s.t.s vomiting.*

Doctor: Can you tell me what the problem is?
Patient: I've got a terrible headache.
Doctor: (1) .... ?
Patient: Just here.
Doctor: (2) .... ?
Patient: Well, it's really bad. And it throbs.
Doctor: Have you had anything like this before?
Patient: Yes, about every three months. I've had them for the last ten years or so.
Doctor: (3) .... ?
Patient: Usually one or two days. This one started yesterday morning.
Doctor: (4) .... ?
Patient: They usually start just before my period. Sometimes if I eat chocolate. I'm not sure.
Doctor: (5) .... ?
Patient: If I lie down in a dark room it helps. Light makes them worse.
Doctor (6) .... ?
Patient: If I move my head, it gets more painful.
Doctor: Apart from the headache, (7) .... ?
Patient: Yes, my eye feels strange. Sometimes I can't see clearly, things get blurred. I feel sick and sometimes I am sick.

47.3 Complete the sentences. Look at Appendix IV on page 144 to help you.

1 Gastric ulcers are associated with a .... , ... pain.
2 Cystitis causes .... , ... pain on passing urine.
3 Patients with a peptic ulcer may say they have a .... , ... pain.
4 Recurrent abdominal pain (RAP) may be described as .... or ....
5 Migraine is often described as a .... pain.
6 People with osteoarthritis often complain of a deep .... centred in the joint.
7 Kidney stone pain is sudden, severe and ....
8 Angina is usually described as a crushing or heavy or .... pain.

Over to you

Choose at least three common conditions and make a note of how patients would describe the pain in English.

48 Taking a history 2

A Drug history

Here is an extract from a medical textbook.

It is essential to obtain full details of all the **drugs** and **medications** taken by the patient. Not infrequently patients forget to mention, or forget the name of, drugs they take. Some may be **over-the-counter remedies** unknown to the general practitioner. The significance of others, such as **herbal remedies** or **laxatives**, may not be appreciated by the patient.

It is necessary to determine the precise identity of the drug, the **dose** used, the **frequency of administration** and the patient's **compliance** or lack of it.

It is important to ask about known drug **allergies** or suspected **drug reactions** and to record the information on the front of the notes to be obvious to any doctor seeing the patient. *Failure to ask the question or to record the answer properly may be lethal.*

To find out about drug history, doctors ask:

Details of drugs and medications

- Are you taking any medication at the moment?
- Which tablet do you take?
- Do you use any over-the-counter remedies or herbal or homeopathic medicines?

Frequency of administration

- How many times a day?

Compliance

- Do you always remember to take it?

Side-effects and allergies

- Do you get any side effects?
- Do you know if you are allergic to any drug?

If the answer is Yes: **What symptoms do you get after taking it?**

B Family history

Note the age, health or **cause of death** of parents, **siblings** (brothers and sisters), **spouse** (husband or wife), and children. To find out about family history, doctors ask:

- Do you have any brothers and sisters?
- Do you have any children?
- Are all your close relatives alive?
- Are your parents alive and well?
- Is anyone taking regular medication?
- How old was he when he died?
- Do you know the cause of death? / What did he die of?
- Does anyone in your family have a serious illness?

C Social and personal history

Record the relevant information about **occupation**, **housing** and **personal habits** including **recreation**, physical exercise, alcohol and tobacco and, in the case of children, about school and family relationships. Typical questions in taking a social and personal history are:

- What kind of house do you live in?
- Do you live alone?
- Who shares your home with you?
- How old are your children?
- Are any of them at nursery or school?
- What's your occupation?
- Do you have any problems at work?
- Do you have any financial problems?
- Do you have any hobbies or interests?
- What about exercise?
- Do you smoke?
- How many a day?
- Have you tried giving up?
- What about alcohol?
- Wine, beer or spirits?
- Can you give up alcohol when you want?
- How much do you drink in a week?
- What's the most you would drink in a week?
- Are you aware of any difference in your alcohol consumption over the past five years?

48.1 Complete the sentences. Look at A, B and C opposite to help you.

1 Pharmacies sell a wide variety of ..........-..........-.......... remedies as well as dispensing prescriptions from physicians.
2 The .... is the quantity of the medication to be taken at any one time.
3 A drug .... is hypersensitivity to a particular drug.
4 A .... is a medication prepared from plants, especially a traditional remedy.
5 Your brothers and your sisters are your ....
6 .... is what you do for physical or mental stimulus outside work.
7 .... can take many forms: apartments, single rooms, houses, hostels.
8 The patient's .... to drug treatment, his willingness or ability to take the right dose at the right time and frequency, is essential.

48.2 Write the doctor's questions. Look at B opposite to help you.

Doctor: (1) .... ?
Patient: My father died twenty years ago but my mother is in good health still. She's seventy now.
Doctor: (2) .... ?
Patient: I was still at school. He was forty-one.
Doctor: (3) .... ?
Patient: He had a heart attack.
Doctor: (4) .... ?
Patient: I've got a sister of forty-five and a brother who's thirty-six.
Doctor: (5) .... ?
Patient: No, I had an elder brother but he died in his forties. He was forty-two.
Doctor: (6) .... ?
Patient: Like my father, a heart attack.
Doctor: (7) .... ?
Patient: Not that I know of.
Doctor: (8) As far as you know .... ?
Patient: Apart from me, no.
Doctor: (9) .... ?
Patient: Yes, a boy and a girl. He's fourteen and she's twelve.

48.3 Study the social history of Mr Black. Write the questions the doctor asked to obtain the numbered information. Look at C opposite to help you.

Social history: Mr G. Black

Home – Lives in a detached house with a large garden(1).

Family – Four children: two girls aged 3 and 4, two boys aged 6 and 8. All are being taught at home by his wife(2).

Occupation – Manager of a DIY warehouse. Stressful job involving dealing with frequent staff problems and meeting monthly sales targets. Large mortgage(3).

Personal interests – Has little time for exercise or interests outside work(4).

Habits – Presently smoking 20 per day(5). Has tried nicotine patches without success(6). Average alcohol intake 3 units per day at weekends(7). No problem with alcohol withdrawal(8).

Over to you

Write a social history of a patient you know. Make a note of the questions you would ask to obtain the information.

49 Taking a history 3

A Reviewing the systems

Once you know the main reason why the patient wants medical attention, it is sensible to ask about the systems to determine the patient's general state of health and to check for any additional problems. The patient should be encouraged to describe symptoms spontaneously. Initial questions should be **open-ended** and as general as possible. Follow up with more specific questions if needed, but avoid putting words in the patient's mouth.

Open-ended questions
What's your appetite like?
How's your vision?

Closed questions
Have you eaten today?
Is your vision ever blurry?

B Asking about the central nervous system

1 **Do you suffer from headaches?**
2 **Have you ever had a blackout?**
3 **What about fits?**
4 **Have you had any dizziness?**
5 **Do you get ringing in the ears?**
6 **Have you ever experienced any numbness or tingling in your hands or feet?**
7 **Do you have any problems sleeping?**

C Patient ideas, concerns and expectations

It is important during the consultation to give patients the chance to express their own ideas and concerns about their problem and to determine what their expectations are. The letters **ICE** (Ideas, Concerns and Expectations) are a way of remembering this. Typical questions are:

Ideas
- **What do you know about this problem/condition/illness?**
- **Do you have any ideas about this?**
- **How do you think you got this problem?**
- **What do you mean by …?**

Concerns
- **What are your worries about this?**
- **Do you have any concerns?**
- **How might this affect the rest of your family?**

Expectations
- **What do you think will happen?**
- **What do you expect from me?**
- **What were you hoping we could do for you?**

D Phrasal verbs in history-taking

Phrasal verbs are often used in informal spoken English. Both patients and doctors may use them in consultations. A phrasal verb may have several meanings according to context.

Phrasal verb	Example	Meaning
bring on	Is there anything special that **brings on** the pain?	cause, induce
bring up	When you cough, do you **bring up** any phlegm?	expectorate, vomit
carry on	**Carry on** taking the painkillers for another week.	continue
come on	When does the pain **come on**?	commence
give up	My advice is to **give up** smoking.	stop
put on	I've **put on** a lot of weight in the last month or so.	gain weight
turn out	She had all the tests and it **turned out** to be cancer.	happen in the end
turn up	The rash just **turned up** out of nowhere.	appear unexpectedly

49.1

Match the numbered questions (1–7) in B opposite to the symptoms for the central nervous system (a–f). There are two questions for one of the symptoms.

a headaches
b hearing symptoms
c faints
d tingling (paraesthesiae)
e fits
f sleep patterns

49.2

Read the extract from a consultation. In the numbered questions (1–4), is the doctor encouraging the patient to talk about her ideas (I), her concerns (C) or her expectations (E)? Look at C opposite to help you.

Patient: I'm a bit concerned about my colic. I had a friend with something similar and it turned out to be more serious. It's got me worried.
Doctor: (1) What do you mean by colic?
Patient: A pain in the stomach.
Doctor: (2) What do you think might have brought this on?
Patient: It just seemed to come on. I don't know what it is.
Doctor: You said you were a bit worried because your friend had a similar problem. (3) What are your worries about this?
Patient: Yes, I had a friend. She turned out to have stomach cancer. She actually died in the end.
Doctor: (4) What were you hoping I could do for you today?
Patient: I just want to know that I don't have anything too serious.

49.3

Complete the sentences with phrasal verbs. Look at D opposite to help you.

1 The headaches .... in the morning.
2 However much I eat, I don't seem to .... any weight.
3 I've tried to .... smoking several times.
4 I'm so depressed I don't feel I can ....
5 When I cough, I .... phlegm.
6 He thought he had stomach ache but it .... to be cancer.

Over to you

Write your own questions about the alimentary system using the checklist. Look at B opposite to help you. Look at Unit 20 if you need more help.

Condition of mouth
Difficulty with swallowing (dysphagia)
Indigestion
Heartburn
Abdominal pain
Weight loss
Change in bowel habit
Colour of motion (e.g. pale, dark, black, fresh blood)

50 Physical examination

A Examining a patient

When examining a patient, you should:

1 Introduce yourself, if necessary.
Good morning, I'm Dr Mason.

2 Brief the patient on what he/she should expect in a clear and simple way.
Do you know what we're going to do this morning?
What we're going to do today is ...
I'm going to examine your ... so I can find out what's causing this ...
What we do is ...
What happens is that ...
I'll ask you to ...
Are you ready?
OK?

3 Instruct the patient in a clear but polite manner (see B below).

4 Show sensitivity to the patient's needs and respond to discomfort, reassuring if appropriate.
You might feel a little bit of discomfort.
This might hurt a little but I'll be quick.
Tell me if it hurts.
Let me know if it's sore.
It will be over very quickly.
It won't take long.
You're doing very well.

5 Talk the patient through the procedure.
I'm just going to ...
First I'll ...
Then I'll ...
Now I'm going to ...
You'll feel ...
When it's over, I'll ...
That's it. All over.

6 Share your findings with the patient.
Well, I'm fairly certain you've got a ...
One possibility is it could be what we call ...
I haven't found anything to suggest any problems.

B Giving instructions

When examining a patient, polite forms are often used for the initial instruction:

Could you bend forward as far as you can?
If you could cross your arms in front of your chest.
What I'd like to do is examine you standing up.

After that direct instructions may be used:

Stand with your feet together.
Lie perfectly still.

To soften an instruction:

Can you just turn to the side again?
Could you just lie on the couch?

For a list of verbs commonly used in instructions during examinations and investigations, see Appendix V on page 145.

50.1 Complete the instructions using suitable verbs. Look at Appendix V on page 145 to help you.

1 .... straight ahead and .... your nose with your right forefinger; then with your left forefinger.
2 .... on the edge of the couch and .... your legs hang loosely.
3 .... your eyelids tightly.
4 .... all your clothes .... down to your underwear.
5 .... your chest with your chin.
6 .... slowly and look over your left shoulder.
7 .... on your side.
8 .... looking at me.
9 .... your sleeve.
10 .... the pin, not the light.

50.2 Write instructions to describe the movements in the pictures. Look at Appendix V on page 145 to help you.

1 2 3 4 5 6 7 8 9 10 11 12 13

50.3 Complete the instructions for a lumbar puncture with words and phrases from A and B opposite.

Morning Mr Maxwell, all right? Now, (1) .... ? Well, (2) .... put a little needle into your spine and take some fluid off your back to find out what's giving you these headaches. You might feel (3) .... but it won't (4) .... When it's (5) .... , we'll ask you to lie still for a few hours. Now Mr Maxwell, (6) .... roll onto your left side? I want you to curl up into a little ball. So could you bend your knees up and tuck your head in for me. That's fine, lovely, OK. Now (7) .... swab down your back with some antiseptic, all right? It'll be a bit cold. Are you (8) .... ? Now I'm going to give you a local anaesthetic so it won't be sore. You'll feel just a slight jab. OK, scratch coming now. There. We'll wait for a few minutes for that to take effect. Right now, lie still, that's very important. Now (9) .... me pressing down as I put the needle in. You're doing (10) .... OK. That's it. All (11) ....

Over to you

Think of an examination you frequently carry out. Write down the instructions you would give the patient.

51 Mental state examination

A Some symptoms of psychiatric disorders

- A **delusion** is a firmly held belief which is wrong but not open to argument. For example, a **deluded** patient may not accept that his psychiatrist is in fact a psychiatrist.
- **Dementia** is significant mental deterioration due to physical changes in the brain.
- **Disorientation** is mental confusion about time, place or identity.
- **Hallucinations** are apparently normal perceptions which happen without the appropriate stimulus. Any of the senses can be involved but especially vision and hearing.
- **Illusions** are misinterpretations of real stimuli.
- **Obsessional** symptoms are stereotyped ideas or impulses which the patient cannot resist. They include obsessional thoughts and obsessional rituals.

B Mood

When describing a patient's mood, it is better for doctors to use the patient's own words rather than their own subjective description of the patient's mood. Patients may say:

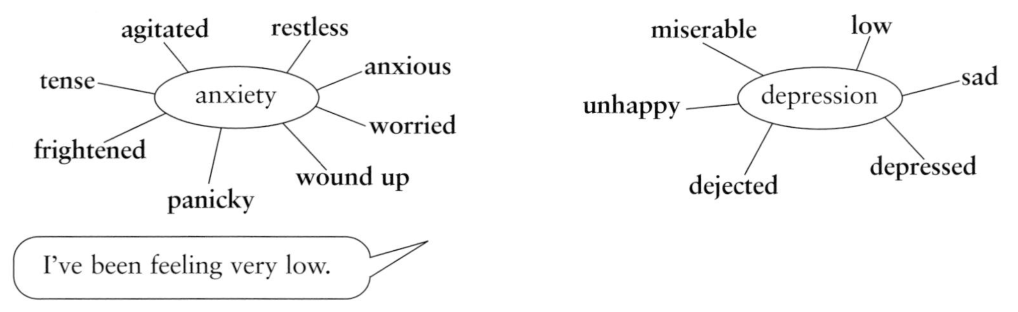

> I've been feeling very low.

Patients may also use the following adjectives to describe their condition:

> I get very **confused** about time. I can't remember what day it is.
>
> People make me **angry**. They're so irritating.
>
> I'm too **embarrassed** by my appearance to go out. I stay at home.
>
> I get **muddled** when I'm shopping. I go out for milk and I come back with cheese.

C Typical questions from a mental state examination

1. **Can you describe your mood at the moment?**
2. **How long have you been feeling like this?**
3. **Do you take pleasure in anything?**
4. **How are your energy levels?**
5. **What's your appetite like?**
6. **Have you noticed any change in your weight?**
7. **How are you sleeping?**
8. **Can you keep your mind on things?**
9. **What do you feel the future holds for you?**
10. **Have you ever felt that you don't want to go on?**
11. **Have you ever thought of suicide?**

51.1 Complete the sentences. Look at A opposite to help you.

1 The patient believes that people can see through walls. He's suffering from a ....
2 The patient sees her long-dead sisters in her garden. She's suffering from ....
3 The patient perceives tree branches as snakes. He's experiencing an ....
4 The patient washes her hands five time before every meal. Her behaviour is ....
5 The patient thinks the nurse is her daughter. She's ....
6 A patient is confused about where she is. She's suffering from ....

51.2 Complete the table with words from A opposite. Then complete the sentences below.

Noun	Adjective
confusion	
	deluded
depression	depressive (illness)
	(patient)
	disoriented
obsession	(symptoms, thoughts)
	obsessive compulsive (disorder)
psychiatry (field) (practitioner)	

1 Patients in .... hospitals receive fewer get-well cards than others.
2 Impaired concentration is a characteristic symptom of ....
3 The potential risk of suicide should always be assessed in the severely ....
4 A .... person may be confused about who they are or where they are.

51.3 Write the doctor's questions in the mental state examination. Look at C opposite to help you.

Doctor: (1) .... ?
Patient: I feel low. I'm not enjoying life.
Doctor: (2) .... ?
Patient: No, nothing.
Doctor: (3) .... ?
Patient: I feel run down. I'm really tired.
Doctor: (4) .... ?
Patient: For months now.
Doctor: (5) .... ?
Patient: I can't get to sleep and when I do sleep I wake up early.
Doctor: (6) .... ?
Patient: I've got no appetite. I don't enjoy food.
Doctor: (7) .... ?
Patient: I'm losing weight.
Doctor: (8) .... ?
Patient: I can't remember where I've put anything.
Doctor: (9) .... ?
Patient: Don't like thinking about it.
Doctor: (10) .... ?
Patient: I've thought about it but I don't have the courage.

Over to you

Describe a psychiatric case you have encountered in your professional career.

52 Explaining diagnosis and management

A Explanations

The final part of a consultation is the explanation which should cover:

1 The **diagnosis** – identification of a disease from its signs and symptoms.

You're suffering from …
You've developed …
You have …
This is (mainly) because …
This is why …

2 The **management plan**, including investigations and treatment.

I'll make you an appointment with …
I'm going to start you on medication to …
I'm going to have you admitted to …
I'll arrange for you to …
You'll be given …
I expect you'll have …
They may advise …

3 General advice about any **change in lifestyle** that may be needed, for example giving up smoking or drinking less alcohol.

The nurse will give you advice on …
You should try to give up …
I want you to …
It's important that you …

4 The **prognosis** – what is likely to happen because of a disease, stressing that nothing is certain.

I expect the treatment will …
Hopefully we can …
We can never be absolutely certain about …
You should remain optimistic.

5 **Question time** – where the patient can ask questions about his/her illness.

Do you have any questions?
Is there anything you'd like to ask?

B Using lay terms in explanations

Explanations should be given in words the patient will understand, avoiding medical jargon. Using **lay terms** – words familiar to people without medical knowledge – can help patients understand explanations. For a list of some common lay terms for conditions, parts of the body and medication, see Appendix VI on page 146.

C An explanation of angina

Having examined you, I'm confident that you're suffering from angina.

The heart is a pump. The more you do physically, the harder it has to work. But as we get older, the blood vessels which supply oxygen to the heart begin to harden and get furred up, so they become narrower. They can't supply all the oxygen the heart needs. The result is the pain you feel as angina.

Because you're experiencing pain at rest as well as on exertion, I'm going to have you admitted to the coronary care unit right away so that your treatment can start at once. You'll be given drugs to ease the pain and I expect you'll have an angiogram. They may advise surgery or angioplasty – that's a way of opening up the blood vessels to the heart so they can provide more oxygen.

You should try to give up smoking. You won't be able to smoke at all in hospital so it's a good time to stop.

I expect the treatment will improve your pain at least and may get rid of it completely. We can never be absolutely certain about the future but you should remain optimistic. Do you have any questions?

52.1 Match the stages of a consultation (1–4) with the sentences used (a–f). Look at A and C opposite to help you.

1 diagnosis
2 management
3 general advice
4 prognosis

a I'm going to have you admitted to the coronary care unit.
b I expect the treatment will improve your pain at least and may get rid of it completely.
c Having examined you, I'm confident that you're suffering from angina.
d You should try to give up smoking.
e We can never be absolutely certain about the future but you should remain optimistic.
f You'll be given drugs to ease the pain and I expect you'll have an angiogram.

52.2 Replace the underlined words and phrases with appropriate lay terms. Look at Appendix VI on page 146 to help you.

1 Mr Harris, I'm afraid your wife's suffered an acute cerebrovascular event.
2 The urethra runs through the middle of the prostate.
3 The reason for your nocturia is that your prostate is enlarged.
4 Do you suffer from dyspnoea when you exert yourself?
5 I'm going to give you an analgesic and an anti-inflammatory for your sprained ankle.
6 I'm putting you on anti-depressants for a short time to help you get back to normal life.

52.3 Complete the explanation of diabetes. Look at A opposite to help you.

You've (1) .... Type 2 diabetes. This is (2) .... very overweight. Your body isn't producing enough insulin. (3) .... you feel so thirsty and why you pass urine so frequently. It's also the reason you have this very itchy rash and you have a problem with your eyes.

The nurse will (4) .... your diet and I'll (5) .... a dietician. I'm (6) .... tablets to control your high blood sugar. You don't need insulin right now but it is possible you might need it in the future.

You (7) .... lose weight and I (8) .... see a podiatrist. It's important with diabetes that you take good care of your feet. You should also see your optician every six months for eye checks.

Diabetes is a serious condition and can affect your heart, blood pressure, circulation, kidneys and vision but we can limit these problems by controlling your blood sugar.

No case of diabetes can be described as mild. I'll (9) .... to attend the diabetic clinic every two months so we can check your progress.

(10) .... reduce this to six monthly visits once your condition is under control.

Is there (11) .... ?

Over to you

Explain a condition of your choice to a patient. Your explanation should cover the points listed in A opposite and use lay terms where possible.

53 Discussing treatment

A Offering options

When discussing options with a patient, doctors may say:

There are a **couple of options** we can use. The **first option** is to try tablets like Prozac that lift you up a bit. The **other option** is counselling.

It can be caused by diet **or** stress. **There are some quite simple tests we can do. If you're still concerned, we can refer you to** a hospital.

B Advising a course of action

When advising a course of action, doctors may say:

Some time off work **might help. If you felt that would be helpful, you could** take a week off and see how you felt after that.

Carry on drinking lots of fluids.

If you still have some pain, you can **keep taking** paracetamol.

Other things might help, like raising the head of your bed. That's one of the simple things we **could start you off with.** You said you haven't tried indigestion remedies. That's **something you could try.**

C Advising patients to avoid something

When advising patients to avoid something, doctors may say:

There are a few things about your lifestyle we could address. Perhaps **cut down on** the amount you're drinking. **Giving up** smoking would help.

Cut out fatty foods.

You **should try to avoid** tight clothing, sitting in deep armchairs and bending, especially after meals.

D Warnings

When a doctor wants to warn a patient that the consequence of ignoring the advice could be serious, he or she may say:

If you aren't feeling better in 7 to 14 days, you **really must come back and see me again.**

If you keep damaging it, you're going to **end up with** a long-term problem.

If you feel that things aren't **settling,** aren't getting back to normal, it's important that you see me again.

It's very important you **don't stop taking** the tablets suddenly or your **symptoms will return.**

53.1 Complete the advice. Look at A, B, C and D opposite to help you.

1 If you .... smoking, you increase the risk of lung cancer and heart disease.
2 Your health would improve if you .... alcohol completely.
3 You could .... with a serious drug problem.
4 I'm going to .... you .... with some tablets. If they don't help, we'll need to think about surgery.
5 .... on the amount of salt you take with your food.
6 I expect things will .... in a few days and you'll be able to get up.
7 Try to .... situations where you feel stressed.

53.2 Match the two parts of the sentences. Look at A, B, C and D opposite to help you.

1 If you still have pain,
2 If you find it difficult to give up smoking,
3 Your symptoms will return
4 If you can't get to sleep,
5 Come back and see me again
6 If you're still concerned,

a try to avoid caffeine later in the day.
b if you find your breathlessness has increased.
c keep taking paracetamol.
d I can refer you to a consultant.
e if you start smoking again.
f I can arrange for nicotine replacement therapy.

53.3 Advise a patient with high blood pressure about physical activity. Look at the information in the table, and at C opposite to help you.

Advice for people with heart disease or high blood pressure	
Do	**Avoid**
Moderate, rhythmic (aerobic) exercise such as brisk walking, cycling or swimming.	Intense exercise such as weight-lifting, press-ups, heavy digging and isometric exercise.
Any regular physical activity that you are used to.	Any sport or activity that brings on angina.
Eat a low cholesterol diet.	Moving from floor to standing exercises too quickly.

You should avoid heavy digging. ....
You shouldn't do any heavy digging. ....
You could try brisk walking. ....

Over to you

A patient of yours, Mrs White, aged 44, has been complaining of very heavy periods. An ultrasound scan of her pelvis shows she has a small fibroid in her uterus. She asks about treatment. These options are possible:

1 Do nothing. The fibroid will shrink when she becomes menopausal.
2 See a gynaecologist who may advise removal of the fibroid or a hysterectomy.

Practise what you would say to Mrs White to advise her of these options.

54 Giving bad news

A Principles of giving bad news

1 Give warning.

I'm afraid your test results aren't very good.
I'm sorry to have to tell you that the news isn't good.

2 Choose an appropriate setting and have a friend or relative of the patient present.
3 Take time.
4 Use appropriate language.
5 Emphasize the positive.

There's still a lot we can do to help you.
Chemotherapy will **make you more comfortable.**

6 Discuss the prognosis.

One can never be certain about these things but I'd say **it's a matter of months rather than years.**

7 Supplement the verbal message.

I'd like to **record** this consultation so you can listen again if anything isn't clear.

8 Arrange a follow-up session.

I'd like to see you again next week.
Can you **come in again** next week?

9 Confirm that the patient understands.

Could you tell me what we're going to do for you?
Is everything clear to you?

B A consultant medical oncologist's report

Mr Harry Scott

Diagnosis: Previous pancreatic cancer

I reviewed Mr Scott in the Oncology Clinic today. He has been less well and has lost 12 kg in the past few months. Unfortunately, his CT scan shows an area of ill-defined low attenuation in the tail of the pancreas. Although this is consistent with focal pancreatitis, the general feeling at the Multidisciplinary Team meeting was that this represents recurrent disease. This is especially likely in view of his clinical deterioration and rising CA19.9. I **discussed** this **with** Mr Scott and his wife. He was obviously **disappointed with** the scan results but still tries to **remain positive**. We discussed the fact that **surgery wasn't an option** and **symptom control** was important. We also discussed the role of palliative Gemcitabine. The potential benefits are small but it is usually **well tolerated** and he was keen to proceed with this. I will therefore **book him into Ward 2** to **start treatment** in the next few weeks and have re-checked his bloods today. In the meantime, I would be very grateful if you would **refer him to** your dietician. He himself is **keen for** this to happen. We will see him back in Clinic once his treatment has started.

54.1 Complete the extract from the oncologist's consultation with Mr Scott. Look at A and B opposite to help you.

Oncologist: Mr Scott, (1) .... so you and Mrs Scott can play back later anything that may not be clear to you today.

(2) .... that the scan results aren't very good. It's likely that you've got a recurrence of cancer in your pancreas. That would explain why you've been feeling so tired, and your loss of appetite and weight.

Mr Scott: Will I need surgery?

Oncologist: Surgery (3) .... at this stage. Although we can't operate, there is still (4) .... You've got tablets for pain relief and we can give you something stronger if you need it. We can also start you on a course of chemotherapy to help with your symptoms. This won't cure you but it will (5) .... It's unusual to have any unpleasant side effects with this kind of chemotherapy. I'd like you too to see a dietician for some advice on what to eat and to help get your appetite back.

Mr Scott: What's my life expectancy? How long have I got?

Oncologist: One can (6) .... People with this condition vary a great deal. I would be wrong to give you a definite time scale but I'd say (7) .... (8) .... all this, but my feeling is it's always best to be honest with people and then you know what's what.

If you're in agreement, I'd like to (9) .... Ward 2 to start your chemo. You'll need to come in every week for the next month.

Is everything clear to you? (10) .... treatment we're going to give you? Are there any particular worries you have?

I'll be seeing you regularly to keep an eye on things so you can ask me any other questions you may have.

54.2 Complete the sentences with words from the box. Look at A and B opposite to help you.

for	into	to	with

1 These results are consistent .... recurrent cancer.
2 His GP referred him .... an oncologist.
3 The patient was disappointed .... the news.
4 The prognosis was discussed .... the patient and his wife.
5 The patient was booked .... the ward for further chemotherapy.
6 He was keen .... this to happen.

Over to you

What would you say to a patient who has a non-malignant brain tumour, about four centimetres in diameter?

55 Data presentation 1

A Referring to a table or figure

In an article, you can write:

Figure 1 Table 1	**shows** X.
X **is shown in**	Figure 1. Table 1.

In a presentation, you can use the same expressions, or you can say:

As you can see in Table 1 …

B Comparing variables

When you refer to a table you will often need to **compare** one variable **with** another:

X was	**twice** **three times**	**as**	effective common	**as**	Y.

Or you may need to compare the same variable at different times:

The number of X in 2000 was	**double** **triple / three times**	**that** in 1990.

There was a	**twofold** **threefold**	**increase in** the number of X **between** 1990 and 2000.

The number of X increased	twofold threefold	between 1990 and 2000.
The number of X	**doubled** **trebled**	

C Approximating

When referring to the data presented on slides, numbers are often **rounded**, for example 41.3 becomes 41 or even 40. When this is done, it is common to use expressions of **approximation**:

Side-effects were reported by	**about** **around** **approximately** **roughly** **some**	forty patients.

Numbers are frequently presented as fractions or percentages, even when the exact number is given.

Roughly two-thirds of patients reported side-effects.

Fifty-five, or **some** two-thirds, of patients reported side-effects.

When you want to emphasize a number, for example 9.8%, you can say:

almost 10%
nearly 10%

more than 9%
over 9%

just under 10%

and when you want to make the number seem small:

less than 10%

55.1 Complete the description of the data in the table. Look at A and B opposite to help you.

Table 1: Incidence of ulcer perforation 1967–1982

		>65	65–74	>75
No. of prescriptions per 1000 (Women)	1967 1982	500 1500		
Perforations (Women)	1967 1982		7 14	10 33
No. of prescriptions (Men)	1967 1982	290 820		
Perforations (Men)	1967 1982		36 28	32 65

Table 1 (1) .... trends in the frequency of hospital admission for perforated peptic ulcer in the United Kingdom (2) .... with changes in the annual prescription rates for non-steroidal anti-inflammatory drugs.
For women over 65 the annual number of prescriptions increased (3) .... from 1967 to 1982, during which rates of perforation of duodenal ulcers (4) .... for those aged 65 to 74 and more than (5) .... for those aged 75 and over. For men over 65, prescriptions showed a similar increase. Although perforation rates were actually lower for those aged 65 to 74 in 1982, there was a (6) .... increase in those aged 75 and above.

(*BMJ* 1986;292: 614 Amended with permission from the BMJ Publishing Group)

55.2 The data in the table is adapted from an article entitled 'Alcohol drinking in middle age'. Choose the correct words to complete the description below. Look at C opposite to help you.

	Alcohol drinking		
Characteristics	**Never (n=300)**	**Infrequent (n=423)**	**Frequent (n=295)**
No. (%) of smokers	54 (18.0)	193 (45.6)	204 (69.2)
Old age measurements (follow up)			
No. (%) who have had myocardial infarction	41 (13.7)	60 (14.2)	54 (18.3)
Results of cognitive assessment			
No. (%) with no impairment (controls)	261 (87)	391 (92.4)	257 (87.1)
No. (%) with mild cognitive impairment	25 (8.3)	15 (3.5)	21 (7.1)
No. (%) with dementia	14 (4.7)	17 (4)	17 (5.8)

Only 54, or (1) .... (under/less) than 20%, of the non-drinkers smoked compared with 204, or (2) .... (almost/over) 70%, of the frequent drinkers. 41, or just (3) .... (less/under) 14% of the non-drinkers had had a myocardial infarction at the end of the follow-up period, compared with 54, or (4) .... (almost/over) 18%, of the frequent drinkers. At the end of the follow-up period, (5) .... (about/over) 90% in all three groups had no cognitive impairment. There was mild impairment in (6) .... (under/around) 8% of the non-drinkers, and in 7% of the frequent drinkers, but in (7) .... (less/under) than 4% of the infrequent drinkers. 14, or (8) .... (approximately/over) 5%, of the non-drinkers had dementia, while 17, or 4% of the infrequent drinkers, and (9) .... (more than/nearly) 6% of the frequent drinkers had dementia.

(*BMJ* 2004;329: 539 Amended with permission from the BMJ Publishing Group)

Over to you

Some people feel that approximating is unscientific. What do you think?

56 Data presentation 2

A Line graphs

Notice the verbs used to describe changes over a period of time.

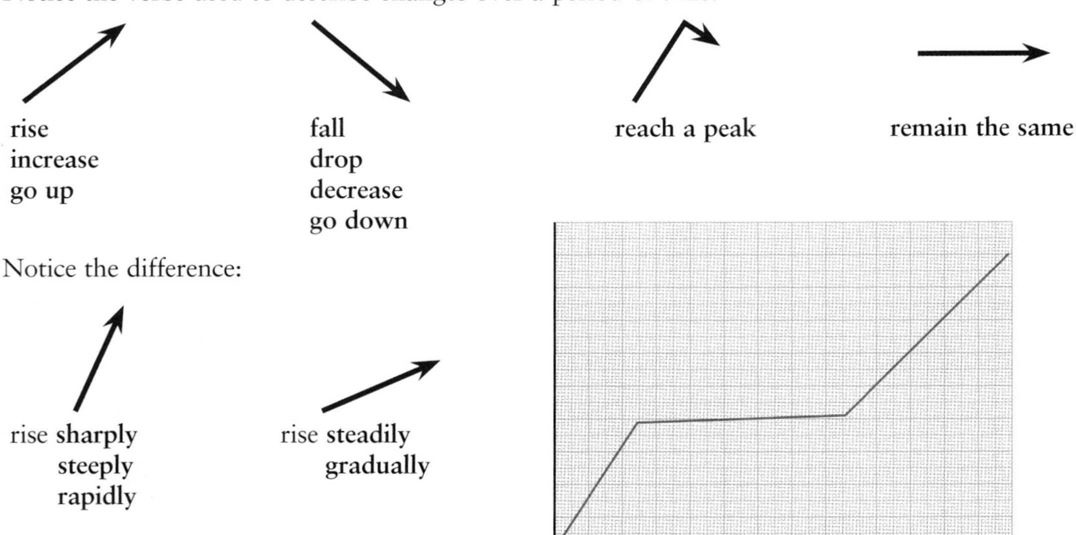

A line graph

B Pie charts

Pie charts are an effective way of showing the relationship of parts to a whole: the complete circle or pie represents the whole, while the parts are represented by **segments** or **slices**. In this pie chart, which shows a health authority's costs, the orange slice **represents** costs of hospital services. So, hospital services **account for** 60% of the costs.

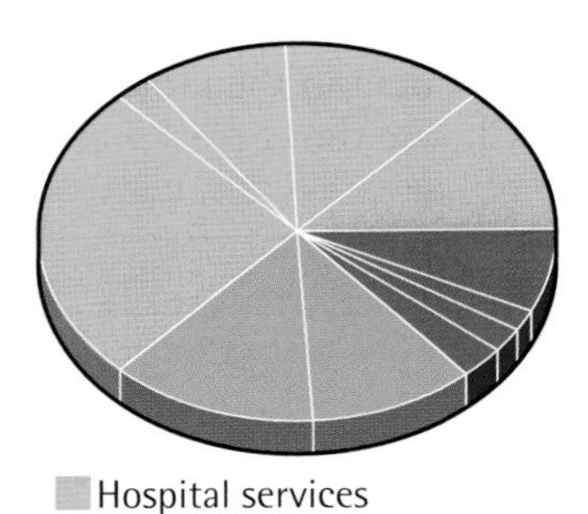

C Describing trends

The **bar chart** below shows the prevalence of HIV in different countries.

HIV rates have **fallen** in some countries as a result of safer sex, but the overall **trend** is an increase in HIV infections. The **steepest** increases in HIV infections occurred in Eastern Europe, Central Asia and East Asia; but sub-Saharan Africa continued to be the most affected part of the world. Adult infection rates in Kenya have **dropped** from a **peak** of 10 per cent in the late 1990s to 7 per cent in 2003. HIV rates in pregnant women in Zimbabwe also **fell** over the past two years.

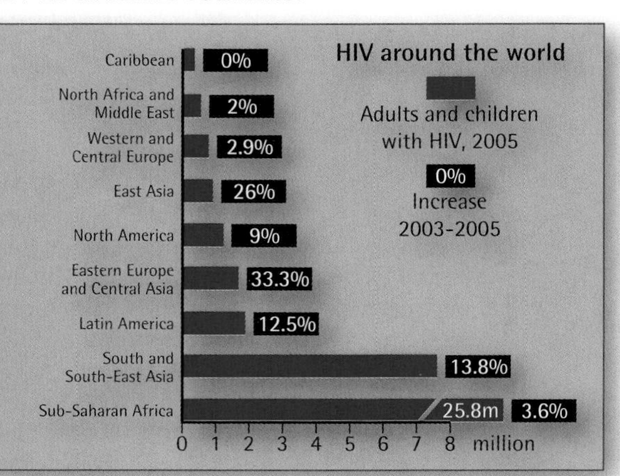

The Times

56.1 Complete the table with words from A opposite. Put a stress mark in front of the stressed syllable of two-syllable words. The first one has been done for you.

Verb	Noun
	'decrease
	drop
	fall
	increase
	rise

56.2 Choose the correct words to complete the description of the bar chart. Look at A and C opposite to help you.

Figure 6 shows that the overall number of cases of HIV (1) .... (dropped / remained the same) yearly until 1988, then (2) .... (rose/dropped) (3) .... (sharply/gradually). Cases in heterosexual men and women (4) .... (increased/fell) (5) .... (steeply/steadily), especially for people exposed abroad. The number of infected people injecting drugs (6) .... (fell/rose) after (7) .... (reaching a peak / dropping) in 1987. This may be because of the development of needle exchange schemes. Mother-to-infant transmission (8) .... (represented/accounted) for a small number of cases. Careful management of labour greatly reduced the number in the late 1990s.

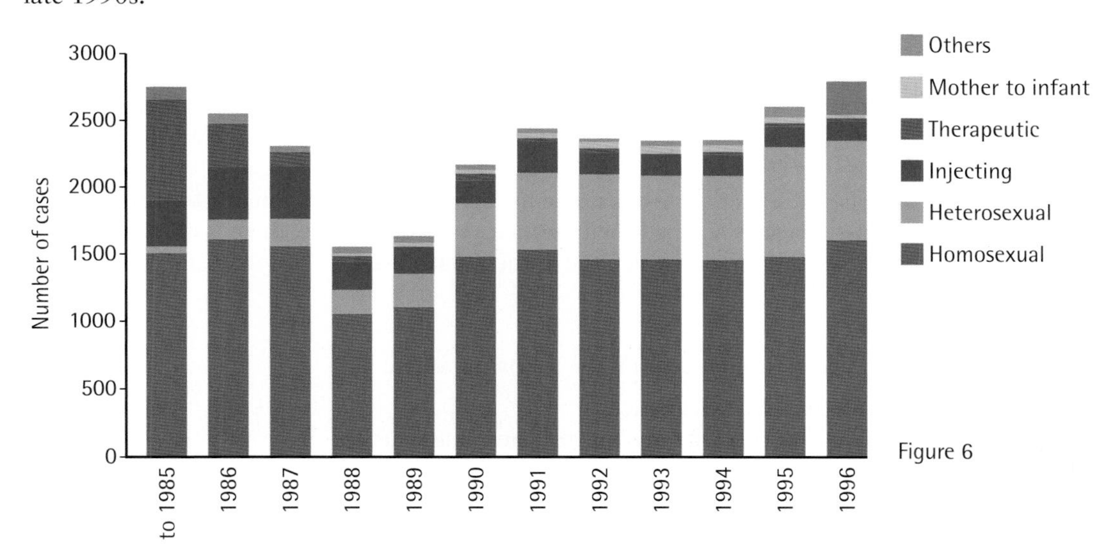

Figure 6

56.3 Which type of graph would best represent the following data?

1 Birth rates in the UK from 1980 to 2010
2 Incapacity due to five different causes from 1995 to 2005
3 Sources of ionizing radiation in the UK

Over to you

Draw a pie chart or bar chart which shows the approximate costs of the health service in your country. Practise describing the chart.

57 Research articles

A The structure of a research article

Research articles are typically divided into four main sections:

Introduction
Methods
Results
Discussion

This is sometimes called the **IMRaD** structure of articles.

The **Introduction** contains **background** information; in other words, it reminds the reader what is already known about the subject. It includes information about previous **studies**, and explains what has not been **investigated** previously. Finally, there is usually a statement of the **objective**, or purpose of the research (why they did it).

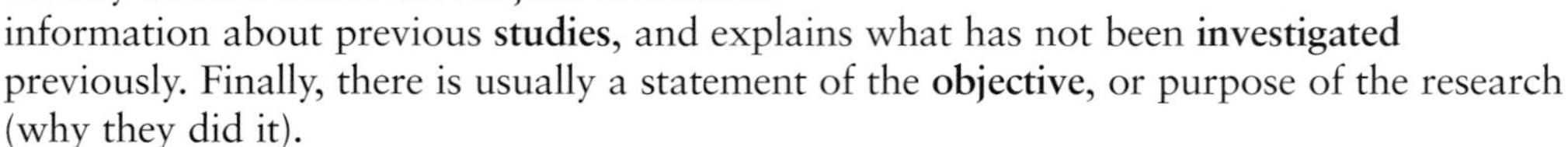

In the case of clinical research, the **Methods** section gives details of the people who were studied – the **participants** in the research. The method section also contains information about any **intervention** carried out, for example medication, advice, operations. It gives details of the **steps taken** in the study, how the participants were chosen, and includes the main things measured, such as blood levels. Finally, there is information about **statistical analysis**.

The **Results** section tells what was found, the **findings** of the study.

The **Discussion** section contains explanations, and claims for the importance of the study. It may also list **limitations**, or parts of the study which were unsatisfactory, and suggest what research needs to be done in the future. There is usually a **Conclusion**, which is sometimes a separate section.

At the end of most articles, there is a short section called **Acknowledgements**. In this the authors thank people who have helped them in their research. Finally, there is a list of **References** – the books and articles which the authors have used.

B Objectives

Statements about objectives often contain the following verbs:

assess	We assessed whether ...
determine	The aim of our study was to determine whether ...
investigate	We investigated the ...
evaluate	This study evaluated the ...

C Main findings

The Discussion section usually begins with a summary of the main findings. This is related to the objective of the study. Typical verbs include:

show	We have shown that ...
confirm	Our study confirmed that ...
provide evidence	These findings provide strong evidence that ...

If the results are less certain:

suggest	These results suggest that ...

and with negative results:

fail to	This study failed to show that ...

57.1 Read the eight extracts from an article in the British Medical Journal entitled 'Paternal age and schizophrenia: a population based cohort study'. Decide which section of the article each extract comes from. There are two extracts from each of the four sections. Look at Unit 46 again if you need more help.

1 People with older fathers were more likely to lose their parents before they reached the age of 18 years.
2 Using a large Swedish record linkage database, we investigated the association between paternal age and schizophrenia in offspring.
3 We used Cox's proportional hazards models to assess the influence of paternal age on psychosis.
4 Our cohort comprised 754,330 people born in Sweden between 1973 and 1980 and still alive and resident in Sweden at the age of 16 years.
5 Our findings confirm an association between increased paternal age and schizophrenia in offspring, which remained even after we controlled for a wide range of potential confounding factors.
6 There is growing evidence that factors operating at different points in life contribute to an individual's risk of developing schizophrenia.
7 The main limitation of our analysis is that case ascertainment was based on people admitted to hospital only with diagnoses recorded on an administrative database.
8 Table 1 shows the characteristics of subjects in relation to the age of their father.

(*BMJ* 2004;329: 1070 Amended with permission from the BMJ Publishing Group)

57.2 Look at the research questions (1–4) and write a statement of the objective of each study, using an appropriate form of the verb in brackets. Look at B opposite to help you.

1 Can calcium and vitamin D supplementation reduce the risk of fractures in postmenopausal women? (assess)
2 Does the way doctors dress influence patients' confidence and trust in them? (determine)
3 Is there a risk of herpes virus 8 (HHV-8) transmission by blood transfusion? (evaluate)
4 Is there an association between never being married and increased risk of death? (investigate)

57.3 Now write a sentence about the main finding in each of the studies in 57.2 above, assuming a result as shown in brackets below. Look at C opposite to help you.

1 (no)
2 (yes)
3 (yes)
4 (uncertain)

Over to you

When you read a research article, which section do you read first? Why?

58 Abstracts

A Structured abstracts

An **abstract** is a type of **summary**, and may be found in special collections of abstracts, such as Medline, or in conference programmes, as well as at the beginning of a research article. Many journals require **contributors** (authors wishing to publish their articles in the journal) to provide a **structured abstract** – an abstract which is divided into specific sections.

B The BMJ abstract

The *British Medical Journal* structured abstract is divided into the following sections:

Objective
Design
Setting
Subjects
Main outcome measure
Results
Conclusion

Here is an example:

AGE AT RETIREMENT AND LONG TERM SURVIVAL OF AN INDUSTRIAL POPULATION: PROSPECTIVE COHORT STUDY

OBJECTIVE
To **assess whether** early retirement is **associated with** better survival.

DESIGN
Long term prospective cohort study.

SETTING
Petroleum and petrochemical industry, United States.

SUBJECTS
Past employees of Shell Oil who retired at ages 55, 60, and 65 between 1 January 1973 and 31 December 2003.

MAIN OUTCOME MEASURE
Hazard ratio of death **adjusted for** sex, year of entry to study, and socioeconomic status.

RESULTS
Subjects who retired early at 55 and who were still alive at 65 had a significantly higher mortality than those who retired at 65 (hazard ratio 1.37, 95% confidence interval 1.09 to 1.73). Mortality was also significantly higher for subjects in the first 10 years after retirement at 55 compared with those who continued working (1.89, 1.58 to 2.27). After adjustment, mortality was similar between those who retired at 60 and those who retired at 65 (1.06, 0.92 to 1.22). Mortality did not differ for the first five years after retirement at 60 compared with continuing work at 60 (1.04, 0.82 to 1.31).

CONCLUSIONS
Retiring early at 55 or 60 was not associated with better survival than retiring at 65 in a cohort of past employees of the petrochemical industry. Mortality was higher in employees who retired at 55 than in those who continued working.

(*BMJ* 2005;331: 995 Amended with permission from the BMJ Publishing Group)

Note: Slightly different headings are used in some journals. For example, *The New England Journal of Medicine* divides articles into Background, Methods, Results and Conclusions. You must consult the Guide to Contributors for the precise requirements of the journal you wish to contribute to.

58.1 Complete the sentences. Look at B opposite to help you.

1 The .... is the aim or purpose of the research.
2 The .... is the location – the country, or part of a country (e.g. a hospital, school, etc.).
3 The .... are the people that the researchers studied.
4 An .... is a result.
5 The .... is the type of study, for example randomized controlled trial.

58.2 Answer the questions about the abstract in B opposite.

1 Who took part in the study?
2 What was the aim of the study?
3 Where was it carried out?
4 What did they measure?
5 What type of study was it?
6 According to this study, does retiring early prolong life?

58.3 The sections of the abstract below are in the wrong order, and the headings have been removed. Decide the correct order, and give each section a title from B opposite.

1 England, Scotland, and Wales.
2 History of asthma, wheezy bronchitis, or wheezing obtained from interview with subjects' parents at ages 7, 11, and 16 and reported at interview by subjects at ages 23 and 33.
3 To describe the incidence [...] of wheezing illness from birth to age 33 and the relation of incidence to perinatal, medical, social, environmental, and lifestyle factors.
4 The cumulative incidence of wheezing illness was 18% by age 7, 24% by age 16, and 43% by age 33. Incidence during childhood was strongly and independently associated with pneumonia, hay fever, and eczema. [...] Incidence from age 17 to 33 was associated strongly with active cigarette smoking and a history of hay fever. [...]
5 Atopy and active cigarette smoking are major influences on the incidence and recurrence of wheezing during adulthood.
6 18,559 people born on 3–9 March 1958. 5801 (31%) contributed information at ages 7, 11, 16, 23, and 33 years.
7 Prospective longitudinal study.

(*BMJ* 2005; Amended with permission from the BMJ Publishing Group)

58.4 Some journals use different headings to those in the *BMJ*. Match the headings (1–5) to the corresponding *BMJ* headings (a–e).

1 Findings
2 Purpose
3 Background
4 Interpretation
5 Participants

a Introduction
b Objective
c Subjects
d Results
e Conclusion

Over to you

Find an article that interests you. Hide the abstract, then try to write the abstract yourself. Compare your version with the real one.

59 Conference presentations

A The structure of a presentation

Conference (or congress) presentations are typically divided into IMRaD sections (see Unit 57).

If the presentation has a different form, the speaker may start by outlining its structure. This helps to orientate the audience:

I'll begin by …
First of all, I'll …
I'll **then** …
Secondly, I'll …
Finally, I'll …

Many speakers like to start a new section with a **signal:**

Now, …
Moving on to X, …
As far as X is concerned, …

An alternative technique is to use a question:

How did we investigate this problem? (to introduce Method)
What did we find? (to introduce Results)
How does this compare with previous studies? (to introduce Discussion)

B The introduction

A formal way of beginning is:
I'd like to present to you the results of our research into …

Many speakers prefer to begin in a less formal way:
When we first began to look into the question of X, we thought …

You may wish to begin with a generalisation or reference to shared knowledge:
It is well known that …
Many studies have shown that …
X has established clearly that …

C Signalling

Other signals that you may wish to give include:

Emphasizing
- **I'd like to emphasize …**

Listing points
- **Firstly, …**
- **Secondly, …**

Referring to slides
- **This slide shows …**

Giving examples
- **For instance …**
- **… such as …**

Contrasting
- **On the other hand …**
- **In contrast …**
- **However, …**

D The conclusion

It is important to end well, for example by summing up the main conclusions.

So, **In conclusion,** **To sum up,** **Finally,**	we can see … I'd like to say end by … these studies show …

59.1

Here are some extracts from a presentation on carbon monoxide poisoning. After announcing his topic, the speaker talked about the pathophysiology of CO poisoning, then the possible sources of the gas, and finally diagnosis and treatment. Put the extracts in the correct order.

1. There is, however, no evidence at all that giving steroids in pharmacological doses is of any proven prophylactic value. ...
2. How do we diagnose it? The early clinical appearances of carbon monoxide poisoning can be very non-specific. ...
3. So, the actual diagnosis of the condition can be very difficult. ...
4. I'd like to tell you about some of our experiences in relation to carbon monoxide poisoning. ...
5. On the slide here, you'll see that there are a variety of sources of carbon monoxide, such as car exhausts, fires, and so on. ...
6. First of all, I want to concentrate on smoke from fires as one of the most important sources of carbon monoxide. ...
7. Now, in relation to the treatment of carbon monoxide poisoning, 100 per cent oxygen administered through a tight-fitting face mask or endotracheal tube is essential. ...
8. I think it's important to emphasize that the presence of cherry-red mucous membranes is a very, very poor sign. ...
9. Now, you will remember that carbon monoxide strongly binds with haemoglobin to produce carboxyhaemoglobin. ...
10. The only accurate way of detecting whether the patient has been exposed to carbon monoxide is to measure carboxyhaemoglobin in, usually, the venous blood. ...

59.2

Write the word or phrase used by the speaker in the presentation in 59.1 above next to its function. Look at B, C and D opposite to help you.

Emphasizing	
Listing	
Exemplifying	
Contrasting	
Summing up	
Changing topic	
Referring to a slide	
Announcing the topic	

Over to you

What makes a good presentation, in your opinion? Think about:

- loudness of voice
- speed of delivery
- use of emphasis and pausing
- eye contact with the audience
- body language – posture and gestures
- visual aids.

60 Case presentations

A Sections of a case presentation

In a **case presentation**, a doctor presents the details of a patient's case to colleagues at a clinical meeting in a hospital. A typical case presentation is divided into the following sections:

Section	Example language
Introduction	I'd like to **present** Mr Simpson,
Patient's age and occupation	a 34-year-old plumber,
Presenting symptom(s) and duration	who **presented with** a one-month history of breathlessness.
Associated symptoms	He also **complained of** ankle swelling which he'd had for two weeks.
Past medical history	There was no **relevant past history**.
Social history	He was married with one son. He smoked 25 cigarettes a day and drank about 50 units of alcohol per week.
Family history	His father died of myocardial infarction at the age of 42. His mother was **alive and well**.
Findings on examination	**On examination**, he was obese ...
Investigation results	We did a chest X-ray which **showed** ...
Diagnosis	So we thought he had ...
Treatment	We gave him intravenous furosemide and ...
Outcome – what happened	He **responded to treatment** and was **discharged home**.

B Bedside presentation

A less formal type of case presentation can take place at a patient's bed, for example during ward rounds, when the house officer presents a new patient to the consultant; or in a **teaching ward round**, when a medical student presents a case to the **tutor**. This type of presentation begins less formally:

This is Mr Simpson. He's 34, and he's a plumber. He **came into hospital** yesterday, sent by his GP. He's complaining of breathlessness, which he's had for one month ...

C Slides

In a formal presentation, the main points are usually summarized on slides.

Mr Simpson 34, plumber
c/o dyspnoea **1/12**; ankle swelling **2/52**
SH married with 1 son
25 cigs/day; 50 units alcohol/week
PH **nil** relevant
FH father **d.** 42 **MI**; mother **a&w**
OE obese; 2 spider naevi on chest
P 110/min reg.
BP 100/60
CXR enlarged heart and bilat. pleural effusions

60.1 Write the abbreviations in words. Look at Look at A and C opposite and at Appendix II on page 131 to help you.

c/o ....
2/52 ....
PH ....
FH ....
MI ....
BP ....
1/12 ....
nil ....
SH ....
a&w ....
OE ....
CXR ....

60.2 Put the sections of a short informal case presentation in the correct order. Look at A opposite to help you.

1 Mr Collins is a 60-year-old security guard.
2 There was no relevant previous medical history.
3 He smokes 20 cigarettes per day and drinks 15–20 units of alcohol each week.
4 On examination, there was marked tenderness around the lower legs above the ankles and knees. There were crackles at the left base posteriorly in the chest. There was nothing else abnormal to find on examination except for clubbing of the fingers.
5 He presented with a six-week history of pain in the legs.
6 Chest X-ray showed consolidation in the left lower lobe. Bronchoscopy and biopsy showed adenocarcinoma of the lung and computed tomography (CT) scan showed that this was not resectable.
7 Treatment with chemotherapy has resulted in temporary improvement in the chest X-ray but the leg pain has continued to prove difficult to control.
8 The pain, which was located around the ankles, had been increasing in intensity and was associated with local tenderness.
9 On routine questioning, he said that he had had a morning cough with small amounts of white sputum for many years. He produced, once, some streaks of blood in the sputum.

60.3 Read the presentation below and make notes for a slide. Look at C opposite to help you.

I'd like to present Mr McNamara who's a 63-year-old taxi driver who presented to the Outpatient Clinic with a three-month history of increasing shortness of breath and ankle swelling. He had a chronic cough with purulent sputum and occasional haemoptysis. Of note in his past medical history was that he'd had a partial gastrectomy in 1980.

On examination, he was pale. He was apyrexial. He had leg oedema, but no clubbing or lymphadenopathy. And examination of his chest was entirely normal. His liver was palpable 5 centimetres below the costal margin, and was smooth and non-tender, and there was also a scar from his previous operation.

Over to you

Make notes about a patient you know and practise presenting him or her.

Appendix I

Parts of the body

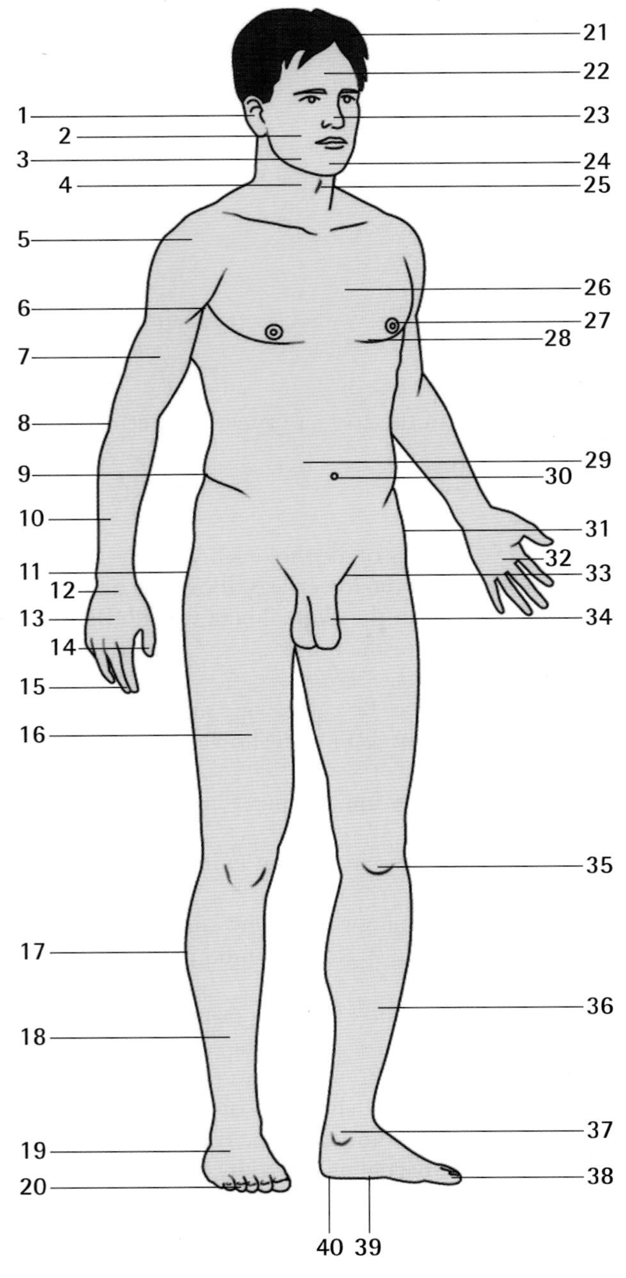

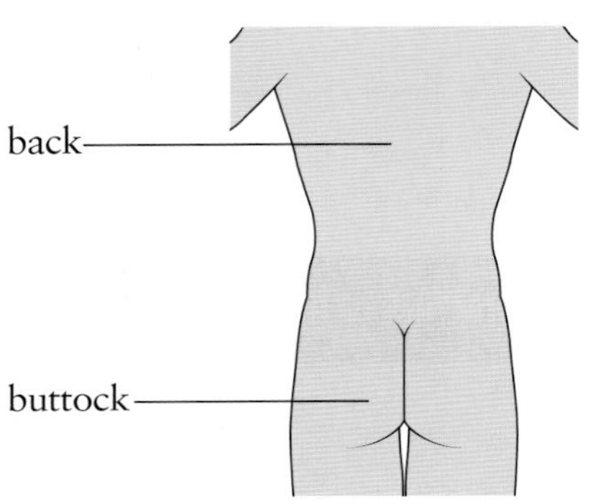

1 ear
2 cheek
3 jaw (mandible)
4 neck
5 shoulder
6 armpit (axilla)
7 upper arm
8 elbow
9 loin
10 forearm
11 buttock
12 wrist
13 hand
14 thumb
15 finger
16 thigh
17 calf
18 leg
19 foot
20 toe
21 hair
22 forehead
23 nose
24 chin
25 Adam's apple (laryngeal prominence)
26 chest (thorax)
27 nipple
28 breast
29 stomach, tummy, belly (abdomen)
30 navel, belly button (umbilicus)
31 hip
32 palm
33 groin (inguinal region)
34 genitals (penis and testicles)
35 knee (patella = kneecap)
36 shin
37 ankle
38 big toe
39 sole
40 heel

Appendix II

Medical abbreviations

Abbreviation or symbol	Meaning
#	fracture
↑	increased/raised
↓	decreased/reduced
♀	female
♂	male
μmol/L	micromols per litre
μg	microgram
1/12	1 month
1/52	1 week
10^9/L	times ten to the power nine per litre [note superscript 9]
A&E	accident and emergency
a&w	alive and well
a.c.	before meals/food (Latin)
a.m.	in the morning (Latin)
A:G	albumen globulin ratio
AB	apex beat
ABC	airways, breathing, circulation
abd / abdo.	abdomen
ACTH	adrenocorticotrophic hormone
AF	atrial fibrillation
AFP	alphafetoprotein
AHA	Area Health Authority
AI	aortic incompetence
AIDS	acquired immunodeficiency disease
AJ	ankle jerk
alk. phos.	alkaline phosphatase
ALT	alanine aminotransferase
AMA	American Medical Association
AN	antenatal
AP	antero-posterior
APH	antepartum haemorrhage
ARM	artificial rupture of membranes
AS	alimentary system
ASD	atrial septal defect
ASO	antistreptolysin O
ATS	antitetanic serum; antitetanus serum
AVF	augmented voltage foot

Abbreviation or symbol	Meaning
AVL	augmented voltage left arm
AVR	augmented voltage right arm
b.d. / b.i.d.	twice a day (Latin)
BAL	blood alcohol level
BB	bed bath; blanket bath
BBB	bundle branch block
BC	bone conduction
BCG	bacille Calmette-Guérin
BF	breast fed
BI	bone injury
BID	brought in dead
BIPP	bismuth iodoform and paraffin paste
BM	bowel movement
BMA	British Medical Association
BMJ	British Medical Journal
BMR	basal metabolic rate
BMSc	Bachelor of Medical Science
BNF	British National Formulary
BNO	bowels not opened
BO	bowels opened
BP	blood pressure
BPC	British Pharmaceutical Codex
BPD	bi-parietal diameter
BS	breath sounds; bowel sounds
BS	Bachelor of Surgery
BWt	birth weight
C	head presentation; centigrade; Celsius
c̄ / c.	with (Latin)
c.c.	with meals/food (Latin)
c/o	complains of
CA / Ca	cancer; carcinoma; calcium
CABG	coronary artery bypass graft
CAD	coronary artery disease
capt.	head presentation
CAT	coaxial or computerised axial tomography
CBT	cognitive behavioural therapy
CCF	congestive cardiac failure

Abbreviation or symbol	Meaning
CCT	Certificate of Completion of Training
CFT	complement fixation test
ChB	Bachelor of Surgery
CHF	chronic heart failure
Chr.CF	chronic cardiac failure
CNS	central nervous system
CO	casualty officer
COAD	chronic obstructive airways disease
COP	change of plaster
CPN	community psychiatric nurse
creps	crepitations
CSF	cerebrospinal fluid
CSSD	Central Sterile Supply Depot
CSU	catheter specimen of urine
CSW	clinical support worker
CT	cerebral tumour; coronary thrombosis; computerized tomography
CV	cardiovascular
CVA	cardiovascular accident; cerebrovascular accident
CVS	cardiovascular system; cerebrovascular system
Cx	cervix
CXR	chest X-ray
D	divorced; died/dead
D&C	dilatation and curettage
D&V	diarrhoea and vomiting
DD	dangerous drugs
DDA	Dangerous Drugs Act
decub.	lying down (Latin)
DIC	drunk in charge
dl	decilitre
DN	District Nurse
DNA	did not attend
DNA	deoxyribonucleic acid
DOA	dead on arrival
DOB	date of birth
DRCOG	Diploma of the Royal College of Obstetricians and Gynaecologists
DRO	Disablement Resettlement Office
DS	disseminated sclerosis

Abbreviation or symbol	Meaning
DTs	delirium tremens (Latin)
DU	duodenal ulcer
DVT	deep venous thrombosis
DWP	Department for Work and Pensions
Dx	diagnosis
E	electrolytes
EBV	Epstein–Barr virus
ECF	extracellular fluid
ECG	electrocardiogram
ECT	electroconvulsive therapy
EDC	expected date of confinement
EDD	expected date of delivery
EDM	early diastolic murmur
EEG	electroencephalogram
ENT	ear, nose and throat
ESN	educationally sub-normal
ESR	erythrocyte sedimentation rate
ETT	exercise tolerance test
EUA	examination under anaesthesia
F	female
fb	finger breadth
FB	foreign body
FBC	full blood count
FH	fetal heart
FH	family history
FHH	fetal heart heard
FHNH	fetal heart not heard
fl / fL	femtolitre
FMFF	fetal movement first felt
FOB	faecal occult blood
FPC	family planning clinic
FRCS	Fellow of the Royal College of Surgeons
FSH	follicle stimulating hormone
FTAT	fluorescent treponemal antibody test
FTBD	fit to be detained; full term born dead
FTND	full term normal delivery
FUO	fever of unknown origin

Abbreviation or symbol	Meaning
FY1	Foundation Year 1
FY2	Foundation Year 2
g	gram
G	gravidity
g/L	grams per litre
GA	general anaesthetic
GB	gall bladder
GC	general condition
GCFT	gonococcal complement fixation test
GCS	Glasgow Coma Scale
GIS	gastro-intestinal system
GMC	General Medical Council
GnRH	gonadotrophin-releasing hormone
GOT	glumatic oxaloacetic transaminase
GP	General Practitioner
GPI	general paralysis of the insane
GPT	glutamic pyruvic transaminase
GTN	glyceryl trinitrate
GTT	glucose tolerance test
GU	gastric ulcer
GUS	genito-urinary system
Gy	Grays
Gyn.	gynaecology
H&P	history and physical examination
Hb / Hgb	haemoglobin
HBP	high blood pressure
HCT / Hct	haematocrit
HDU	high dependency unit
HHV-8	(human) herpes virus 8
HIB	Haemophilus influenzae B
HIV	human immunodeficiency virus
HO	house officer
HP	house physician
HPV	human papilloma virus
HR	heart rate
HS	heart sounds
HVS	high vaginal swab

Abbreviation or symbol	Meaning
IBS	irritable bowel syndrome
ICF	intracellular fluid
ICS	intercostal space
ICU	intensive care unit
ID	infectious disease
Ig	immune globulin
i.m. / IM	intramuscular
IMRaD	introduction, method, results and discussion
infus^n	infusion
IOFB	intra-ocular foreign body
IP	in-patient; interphalangeal
IQ	intelligence quotient
ISQ	condition unchanged / in statu quo (Latin)
IU	international unit
IUD	intrauterine (contraceptive) device
i.v. / IV	intravenous
IVC	inferior vena cava
IVF	in vitro fertilization
IVP	intravenous pyelogram
IVU	intravenous urogram
Ix	investigation
IZS	insulin zinc suspension
JAMA	Journal of the American Medical Association
JVP	jugular venous pressure
K	potassium
KUB	kidney, ureter and bladder
L	left
L/L	litres per litre
LA	left atrium; local anaesthetic
LAD	left axis deviation; left anterior descending
LBP	low back pain; low blood pressure
LDH	lactic dehydrogenase
LE	lupus erythematosus
LFT	liver function test
LH	luteinizing hormone
LHA	Local Health Authority
LIF	left iliac fossa

Abbreviation or symbol	Meaning
LIH	left inguinal hernia
LKS	liver, kidney and spleen
LLL	left lower lobe
LLQ	left lower quadrant
LMN	lower motor neurone
LMP	last menstrual period; left mento-posterior position of fetus
LOA	left occipito-anterior position of fetus
LOP	left occipito-posterior position of fetus
LP	lumbar puncture
LSCS	lower segment caesarean section
LUA	left upper arm
LUQ	left upper quadrant
LV	left ventricle; lumbar vertebra
LVD	left ventricular dysfunction
LVE	left ventricular enlargement
LVF	left ventricular failure
LVH	left ventricular hypertrophy
M	male
M/F	male/female
m/r	modified release
M/W/S	married/widow(er)/single
mane	in the morning (Latin)
MB	Bachelor of Medicine
MCD	mean corpuscular diameter
mcg	microgram
MCH	mean corpuscular haemoglobin
MCHC	mean corpuscular haemoglobin concentration
MCL	mid-clavicular line
MCP	medical care practitioner
MCV	mean corpuscular volume
MD	Doctor of Medicine
MDM	mid-diastolic murmur
mg	milligram
MI	mitral incompetence/insufficiency; myocardial infarction
mitte	give/send (Latin)
ml	millilitre
mmol/L	millimols per litre

Abbreviation or symbol	Meaning
MMR	mass miniature radiography; measles, mumps & rubella
MO	Medical Officer
MOH	Medical Officer of Health
MOP	medical out-patient
MRC	Medical Research Council
MRCP	Member of the Royal College of Physicians
MRI	magnetic resonance imaging
MS	mitral stenosis; multiple sclerosis; musculoskeletal
MSSU	mid-stream specimen of urine
MSU	mid-stream urine
MSW	Medical Social Worker
MVP	mitral valve prolapse
NA	not applicable
Na	sodium
NAD	no abnormality detected
NBI	no bone injury
ND	normal delivery
NE	not engaged
NHS	National Health Service
NIC	National Insurance Certificate
NMC	Nursing and Midwifery Council
NND	neo-natal death
nocte	at night (Latin)
NOF	neck of femur
NP	not palpable; nasal passage
NPO	nothing by mouth (Latin)
NPU	not passed urine
NS	nervous system
NSA	no significant abnormality
NSPCC	National Society for the Prevention of Cruelty to Children
NYD	not yet diagnosed
o.d.	daily (Latin)
O/E	on examination
OA	on admission; osteo-arthritis
OAP	old age pensioner
Obs.	obstetrics
OBS	organic brain syndrome

Abbreviation or symbol	Meaning
oed.	oedema
OM	otitis media
OOH	out of hours
OPD	outpatient department
OSCE	Objective Structured Clinical Examination
OT	operating theatre
OT	occupational therapist
P	pulse; protein; parity
p.c.	after meals/food (Latin)
p.m.	in the afternoon (Latin)
p.o.	by mouth (Latin)
p.r.	by rectum (Latin)
p.r.n.	as required
p.v.	by vagina (Latin)
PA	pernicious anaemia
Para. 2 + 1	full term pregnancies 2, abortions 1
PAT	paroxysmal atrial tachycardia
PBI	protein bound iodine
PBL	problem-based learning
PDA	patent ductus arteriosus
PERLA	pupils equal and reactive to light and accommodation
PET	pre-eclamptic toxaemia
PH	past history
PID	prolapsed intervertebral disc; pelvic inflammatory disease
Pl.	plasma
PLAB	Professional and Linguistic Assessments Board
PM	postmortem
PMB	postmenopausal bleeding
PMH	past medical history
PN	postnatal
PND	postnatal depression; paroxysmal nocturnal dyspnoea
PO_2	pressure of oxygen
POP	plaster of Paris
PPH	postpartum haemorrhage
PRHO	provisionally-registered house officer
PROM	premature rupture of membranes
PSW	Psychiatric Social Worker

Abbreviation or symbol	Meaning
PU	passed urine; peptic ulcer
PUO	pyrexia of unknown or uncertain origin
PVT	paroxysmal ventricular tachycardia
PZI	protamine zinc insulin
q.d.s. / q.i.d.	four times a day (Latin)
R	right; respiration
RA	rheumatoid arthritis; right atrium
RAD	right axis deviation
RBC	red blood cell (count); red blood corpuscles
RBS	random blood sugar
RCA	right coronary artery
ref.	refer
reg.	regular
RGN	Registered General Nurse
Rh.	Rhesus factor; rheumatism
RHA	Regional Health Authority
RI	respiratory infection
RIF	right iliac fossa
RIH	right inguinal hernia
RLL	right lower lobe
RLQ	right lower quadrant
RM	registered midwife
RMO	Regional or Resident Medical Officer
RN	registered nurse
ROA	right occipital anterior
ROM	range of motion
ROP	right occipital posterior
RS	respiratory system
RTA	road traffic accident
RTC	return to clinic
RTI	respiratory tract infection
RUA	right upper arm
RUQ	right upper quadrant
RVE	right ventricular enlargement
RVH	right ventricular hypertrophy
Rx	take (in prescriptions); treatment (in case notes) (Latin)
S	single; sugar

Abbreviation or symbol	Meaning
s.c.	subcutaneous
s.l.	sublingual
SAH	subarachnoidal haemorrhage
SB	still-born
SBE	sub-acute bacterial endocarditis
sep.	separated
SG	specific gravity
SH	social history
SHO	Senior House Officer
SI	sacro-iliac
sig.	write/label (in prescriptions) (Latin)
sl.	slight
SM	systolic murmur
SMR	sub-mucous resection
SN	student nurse
SOB	short of breath
SOBOE	short of breath on exertion
SOP	surgical out-patients
SpR	specialist registrar
SRN	State Registered Nurse
SROM	spontaneous rupture of membranes
stat.	immediately (Latin)
STs	sanitary towels
SVC	superior vena cava
SVD	spontaneous vertex delivery; spontaneous vaginal delivery
sw / sw.	swab
SWD	short wave diathermy
T	temperature
T&A	tonsils and adenoids
t.d.s. / t.i.d.	three times a day (Latin)
T_3	tri-iodothyronine
T_4	tetra-iodothyronine
tabs	tablets
TB	tuberculosis
TI	tricuspid incompetence
TIA	transient ischaemic attack
TM	transport medium

Abbreviation or symbol	Meaning
TMJ	temporo-mandibular joint
TNS	transcutaneous nerve stimulator
TOP	termination of pregnancy
TPHA	treponema pallidum haemagglutination
TPR	temperature, pulse, respiration
TR	temporary resident
TRH	thyrotrophin-releasing hormone
TS	tricuspid stenosis
TSH	thyroid-stimulating hormone
TT	tetanus toxoid; tuberculin tested
TURP	transurethral prostate resection
TV	trichomonas vaginalis
U	urea; unit
U&E	urea and electrolytes
U/L	units per litre
UGS	urogenital system
UMN	upper motor neurone
URTI	upper respiratory tract infection
USS	ultrasound scan
UVL	ultra-violet light
VD	venereal disease
VDRL	venereal disease research laboratory
VE	vaginal examination
VI	virgo intacta
VP	venous pressure
VSD	ventricular septal defect
VV	varicose vein(s)
Vx	vertex
W	widow/widower
WBC	white blood cell count; white blood corpuscles
WCC	white cell count
WHO	World Health Organization
WNL	within normal limits
WR	Wassermann reaction
XR	X-ray
YOB	year of birth

Types of medication

capsules

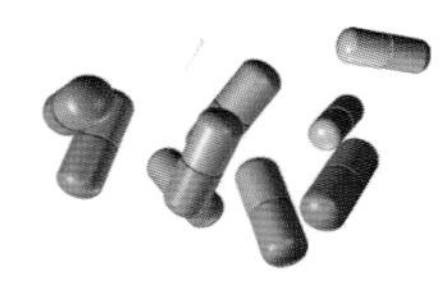

injection

ointment

paste

pessary

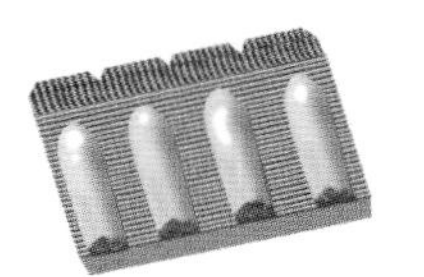

powder

solution

spray

suppository

syrup

tablets

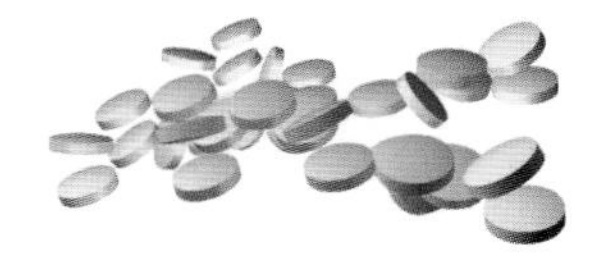

inhaler

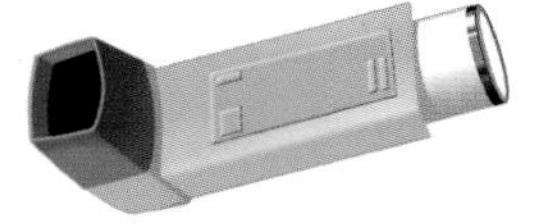

Ointments are greasier than creams and have a thicker texture. This makes them adhere to the affected area longer. **Creams** are more acceptable cosmetically and tend to be used on the face as they are less visible. **Pastes** are stiffer preparations which contain more powdered solids. **Lotions** are liquid and used in areas such as the scalp where an ointment or cream would adhere to the hair.

Appendix IV

Symptoms and pain

Asking about symptoms

Pain is one of the commonest symptoms.

For headaches, a doctor would expect to establish most of the features below. Similar questions can be used for other forms of pain.

Feature	Typical question
Main Site	Where does it hurt? Show me where it hurts.
Radiation	Does it go anywhere else?
Character	Can you describe the pain?
Precipitating factors	Does anything bring them on?
Time of onset	When do they start?
Time of resolution	When do they stop?
Frequency	How often do you get them?
Aggravating factors	Does anything make them worse? Is there anything else that affects them?
Relieving factors	Does anything make them better?
Associated features	Do you feel anything else wrong when it's there? Have you any other problems related to the pain?
Duration	How long do they last?
Severity	How bad is it?

Description of pain

Patient's description of pain	Explanation
aching / an ache	a general pain, often in muscles and joints
boring	like a drill
burning	with heat
colicky	an intermittent pain which varies in intensity, comes and goes in waves
crampy/cramp	an involuntary spasmodic muscle contraction
crushing	a feeling of pressure
dull	a background pain, opposite of sharp
gnawing	biting
gripping	a feeling of tightness
scalding	like boiling water
sharp	acute
stabbing	like a knife
stinging	sharp, burning, like an insect sting
throbbing	with a pulse or beat

Verbs used in instructions

- **bend down**
- **breathe in**
breathe out
- **close** your eyes
- **curl up**
- **do this**
- **follow** my fingertip with your eyes
- **keep** your knee straight
- **let** your wrist go floppy
- **lie** on your side/back
lie on the bed/couch
lie down
- **look** straight ahead
look at something
- **open** your mouth
- **point** to the finger that moves
- **pull** as hard as you can
- **push** as hard as you can
- **put** your head down
put out your tongue
- **raise** your leg
- **roll** on to your back/front
roll over
roll up your sleeve
- **sit**
sit up
- **slide** your hand down your side
- **slip off** your coat
- **stand** straight
stand up
- **take off** your top things
- **tilt** your head back
- **touch** your shoulder with your chin
- **turn** your head to the left
turn on your side

Other instructions:

relax
show me what movements you can manage
tell me if it hurts

Appendix VI

Lay terms and definitions

Explanations should be given in words the patient will understand, avoiding medical jargon. Using **lay terms** – words familiar to people without medical knowledge – can help patients understand explanations.

Some lay terms for medical conditions

Medical conditions	Lay term
acute cerebrovascular event	**stroke**
arrhythmia	**palpitations**
dyspnoea	**breathlessness**
fractured neck of femur	**broken hip**
haematemesis	**vomiting blood**
haematuria	**blood in the urine**
insomnia	**trouble with sleeping**
intermittent claudication	**pains in the back of the legs when walking**
myocardial infarction	**heart attack**
nocturia	**needing to pass urine (water) at night**

Some lay terms for medication

Medical conditions	Lay term
analgesics	**pain killers**
anti-depressants	**tablets to improve your mood**
anti-inflammatories	**medicine to reduce swelling**
broncho-dilator	**a substance which causes the airways to open up**
DMARDs (disease modifying anti-rheumatic drugs)	**pills that help stop arthritis progressing**
diuretics	**water tablets**
hypertension medication	**pills for blood pressure**
hypnotics	**sleeping tablets**
oral contraceptives	**the pill**

Simple definitions

Most patients do not have any medical knowledge, so it is important to use simple words they will understand when talking about certain parts of the body or medical conditions.

Medical term	Simple definition
arteries	tubes which carry blood around the body
benign	not due to cancer or infection
bronchi	airways that connect your windpipe to your lungs
cholesterol	fat that clogs the arteries
intervertebral disks	shock absorbers which separate the bones in your back
oesophagus	the tube that connects the back of the throat to the stomach
pancreas	a gland that helps digestion and makes insulin to control blood sugar
thyroid	a gland that produces some of the hormones required in daily life
urethra	the tube that carries urine from the bladder

Answer key

1.1

Noun	Adjective
fitness	fit
health	healthy
illness	ill
sickness	sick

1.2 complete remission
feel sick
get over
poor health
travel sickness

1.3 1 well 2 unwell/ill/poorly/sick 3 health 4 fit 5 well 6 sick 7 sick 8 illnesses

1.4 1 deteriorated 2 recovered 3 sickness 4 recovery 5 good 6 got over 7 unhealthy

2.1

Anatomical term	Common word
abdomen	stomach, tummy
axilla	armpit
carpus	wrist
coxa	hip
cubitus	elbow
mamma	breast
nates	buttocks
patella	kneecap

2.2 a 1 chest 2 into 3 neck/jaw 4 shoulder 5 down 6 arm
b 1 in 2 groin 3 down

2.3 1 groin 2 tummy/stomach 3 navel 4 chest 5 armpit 6 jaw

2.4

Trunk	Upper limb	Lower limb
abdomen chest loin	elbow finger wrist arm	knee leg thigh

3.1 1 windpipe 2 lung 3 lobes 4 diaphragm 5 airways 6 heart 7 bronchioles

3.2 1e, 2f, 3d, 4g, 5a, 6c, 7b, 8h

3.3 1 organs
2 liver
3 gall bladder
4 kidney
5 kidney
6 spleen
7 bladder

Over to you – sample answer
The spleen is situated on the upper left side of your abdomen, under your ribs. It acts as a filter, helping to destroy old, worn-out blood cells. It also produces cells that help protect your body from infection.

The pancreas is a gland deep inside your abdomen, behind your liver. It normally produces insulin, which your body needs to help it store and utilize glucose, and so it is important in controlling the amount of sugar in your blood.

4.1 1b, 2d, 3e, 4c, 5a

4.2 1 numbness 2 swallowing 3 passing urine/water 4 sweating 5 shake 6 breathing

4.3 1 chew, swallow 2 pass 3 appetite 4 have 5 sense 6 breath

Over to you – sample answer
1 Do you pass water more than usual?
2 Do you drink more than usual?
3 What's your appetite like?
4 Is your sight/vision normal?
5 Have you noticed any numbness in your hands or feet?

5.1
1 A dermatologist specializes / is a specialist in diseases of the skin.
2 A rheumatologist specializes / is a specialist in diseases of the joints.
3 A traumatologist specializes / is a specialist in accident and emergency medicine.
4 A paediatrician specializes / is a specialist in diseases affecting children.
5 An obstetrician specializes / is a specialist in managing pregnancies.

5.2

Verb	Noun (person)	Noun (activity) or thing
'specialize	'specialist	'specialty
'practise	prac'titioner	'practice
con'sult	con'sultant	consul'tation
a'ssist	a'ssistant	a'ssistance
'graduate /'grædʒueɪt/	'graduate /'grædʒuət/	gradu'ation
'qualify		qualifi'cation

5.3 1 specializes in 2 working with 3 interested in 4 good at 5 good with

5.4
consulting rooms
general practitioner
general practice
group practice
health centre
internal medicine
surgical specialties

Over to you – sample answer
To be a surgeon you need to be:
good with your hands
interested in anatomy
able to repeat the same thing without getting bored
able to make decisions fast

6.1 associate specialist
house officer
on call
provisional diagnosis
radio pager
ward round

6.2 1c, 2d, 3a, 4e, 5b

6.3
1 false – After a year, he or she becomes a registered medical practitioner
2 true
3 false – the old terms senior house officer and SHO are still used
4 false – they are usually seen first by one of the junior doctors
5 false – Junior doctors now normally work in shifts ... for example 7 am to 3 pm ... or 11 pm to 7 am. The alternative system is to work from 9 am to 5 pm every day and to take turns to be on call ...

6.4 1 shift 2 admissions 3 clerk 4 round 5 tests 6 discharged 7 training/Foundation

7.1
1 midwife
2 clinical support worker
3 nursing auxiliary
4 staff
5 health visitor
6 charge
7 district
8 ward clerk

7.2 carry out a procedure
change a dressing
check the temperature
give an injection
remove sutures

7.3 1 to perform 2 was performed 3 be performed 4 be performed 5 performed

7.4
1 to carry out
2 have been carried out
3 have been carried out
4 was carried out
5 be carried out

8.1 ambulance technician
artificial limb
club foot
contact lens
health professional
intraocular pressure
occupational therapist
social worker

8.2
1 social worker
2 physiotherapist
3 optician
4 prosthetist / occupational therapist
5 social worker
6 chiropodist/podiatrist

8.3
1 amputation
2 limb
3 therapists
4 rehabilitation
5 splints
6 deformity/deformities
7 relieve

9.1

Verb	Noun
ad'mit	ad'mission
a'ssess	a'ssessment
dis'charge	'discharge
'operate	oper'ation
re'fer	re'ferral
'treat	'treatment

9.2 acutely ill
assessment unit
on duty
referral letter
waiting list

9.3 1 Intensive Care Unit (ICU) / assessment unit
2 High Dependency Unit (HDU)
3 Day Surgery Unit
4 Gynaecology (Emergency)
5 Accident & Emergency (A&E)

9.4 1 clinic/hospital 2 treatment 3 tests 4 department 5 referred 6 admitted 7 day 8 discharged

10.1 1 prescription charges 2 housebound 3 appointment 4 locum 5 out-of-hours

10.2 change dressings
make appointments
make home visits
perform minor surgery
refer a patient
run a clinic
supervise staff
take messages

10.3 1 midwife 2 physiotherapist 3 receptionist 4 practice nurse 5 district nurse 6 health visitor 7 practice manager

10.4 1 surgery 2 appointments 3 Reception 4 visits 5 prescription 6 referrals 7 OOH

11.1 1 undergraduate 2 postgraduate 3 foundation programme 4 undergraduate 5 continuing professional development

11.2 1 elective 2 problem-based learning 3 lectures 4 clinical attachments 5 student selected modules 6 cadavers 7 medical school 8 consultant 9 College 10 seminar

11.3 1 overseas elective 2 problem-based learning / PBL 3 dissection 4 clinical attachment 5 clinical skills 6 student selected module

12.1 1 demonstrate 2 supervise 3 deliver/provide 4 take 5 assess/demonstrate

12.2 1d, 2h, 3f, 4b, 5g, 6c, 7a, 8e

12.3 1 Bachelor of Medicine, Bachelor of Surgery, Fellow of the Royal College of Surgeons of England, Fellow of the Royal College of Surgeons of Ireland
2 Bachelor of Medical Sciences, Doctor of Medicine, Member of the Royal College of Physicians
3 Bachelor of Medicine, Bachelor of Surgery, Fellow of the Royal College of Surgeons
4 Bachelor of Medicine, Doctor of Medicine, Fellow of the Royal College of Physicians

13.1 1 provisional registration 2 limited registration 3 full registration 4 specialist registration

13.2 1 General Medical Council
2 Professional and Linguistic Assessments Board
3 Objective Structured Clinical Examination
4 airways, breathing, circulation
5 Certificate of Completion of Training

13.3 1 limited registration
2 GMC
3 PLAB
4 stations
5 full registration
6 CCT / Certificate of Completion of Training
7 specialist registration

14.1

Noun	Adjective
ex'haustion	ex'hausted
fa'tigue	fa'tigued
'lethargy	le'thargic
'tiredness	'tired

14.2 complain of
off-colour
out of sorts
present with
put on
worn out

14.3 1 presented 2 presentation 3 present 4 presents 5 presenting 6 presentation 7 presenting 8 presentation

14.4 1 complained 2 malaise 3 gained 4 presented 5 constipated

Over to you – sample answer
A 60-year-old man presented to his GP with malaise and fatigue for several months. He also complained of anorexia and weight loss of 10 kg.

15.1 adequate – inadequate
against – for
insidious – sudden
likely – unlikely
severe – mild
rare – common

15.2 bone marrow
differential diagnosis
insidious onset
iron deficiency
pernicious anaemia
progressively increasing
vibration sense

15.3 1 progressive 2 deficiency 3 platelet 4 exclude 5 jaundice 6 palpable 7 adequate 8 breakdown

15.4 1 insidious onset 2 pallor 3 symmetrical 4 vibration 5 sore

Over to you – sample answer
Causes of anaemia mentioned: chronic blood loss associated with carcinoma of the bowel or chronic bleeding ulcer, leukaemia, aplastic anaemia, iron deficiency, pernicious anaemia

16.1

1 skull
2 jaw bone
3 spine
4 breastbone
5 rib
6 collarbone
7 shoulder blade
8 thigh bone
9 kneecap
10 shinbone

16.2 1a, 2c, 3d, 4b, 5e

16.3 1 Reducing 2 unite 3 malunion

17.1

Verb	Noun(s)	Adjective(s)
delay	delay	delayed
develop	development	developed, developing
distend	distension	distended
fail	failure	failing
nourish	nourishment, nutrition	nourished

1 failure 2 distension 3 development 4 nutrition 5 Delay

17.2 1 crawl 2 weaned 3 stature 4 milestones 5 puberty/maturity

17.3 1 mumps 2 lockjaw 3 German measles 4 whooping cough 5 chickenpox 6 measles 7 scarlet fever 8 rheumatic fever 9 polio 10 croup

18.1

Verb	Noun
in'hibit	inhi'bition
pro'duce	pro'duction
re'lease	re'lease
re'place	re'placement
se'crete	se'cretion
'stimulate	stimu'lation

18.2 Answers given are the words used in the original text; possible alternatives are shown in brackets.

1 stimulate (trigger)
2 release (secretion)
3 stimulates (triggers)
4 production (secretion)
5 feeds
6 inhibit
7 secretion (production)
8 stimulates (triggers)
9 produce
10 secretion (production)

18.3 1 diffuse 2 localized 3 deficient 4 Excess

18.4 1d, 2c, 3a, 4e, 5g, 6b, 7f

Over to you – sample answer
Dear Doctor
I would be grateful if you would see this 60-year-old woman who complains of tiredness and constipation. She has gained 5 kilos in the past three months although she says her appetite is poor. She has noticed that her hair has begun to fall out and that her skin is very dry. I am wondering if she is hypothyroid.

19.1

Verb	Noun	Adjective
accommodate	accommodation	
constrict	constriction	constricted
converge	convergence	
dilate	dilation, dilatation	dilated
droop		drooping
oscillate	oscillation	
react	reaction	

19.2 1e, 2c, 3b, 4f, 5d, 6a

19.3 1 pupil 2 flame-shaped 3 cottonwool 4 arteries/arterioles 5 nipping

20.1 1 distended 2 tenderness 3 quadrant 4 guarding 5 rigidity 6 tenderness 7 masses 8 Bowel

20.2 1c, 2e, 3f, 4a, 5b, 6d

20.3 1f, 2a, 3b, 4g, 5e, 6c, 7d

21.1
1 (surgical) removal of the womb
2 heavy periods
3 inflammation of one of the tubes that connect the ovary to the womb
4 biopsy of (small piece of tissue removed from) the neck of the womb

21.2
menarche: 12 yrs
menstrual cycle: $\frac{4\text{–}5}{28}$
LMP: 1/52 ago
menorrhagia? no
dysmenorrhoea? no
discharge? a little white discharge

21.3
menarche: How old were you when you started to get them / your periods?
menstrual cycle: Are your periods regular? How long do the periods last usually?
LMP: When was your last period?
menorrhagia? Would you say they are light or heavy? Do you get clots?
dysmenorrhoea? Do you get period pains?
discharge? Is there any discharge between the periods? What colour is it?

21.4 1 prolonged/heavy 2 clots 3 periods 4 periods 5 flushes 6 oral 7 pill 8 coil

22.1 1 palpitations 2 shortness 3 breath 4 pillows 5 mitral valve

22.2 at rest
atrial fibrillation
cardiac output
cardiac failure
heart failure
on effort
pitting oedema
premature beats

22.3
1 breathlessness / shortness of breath
2 palpitations
3 breathlessness when lying flat
4 swelling

22.4 1 exertion/exercise/effort 2 episodes/attacks 3 palpitations 4 shortness 5 swelling

23.1

Verb	Noun
'auscultate	auscul'tation
e'xamine	exami'nation
in'spect	in'spection
pal'pate	pal'pation
per'cuss	per'cussion

23.2
d Look for clubbing. (inspection)
c Feel the radial pulse. (palpation)
e Locate the apex beat. (palpation)
f Note any thrills. (palpation)
a Measure the heart size. (percussion)
b Are there any murmurs? (auscultation)

23.3 1 force 3 central/peripheral 5 beat 7 murmurs
2 pulses 4 peripheral/central 6 space 8 rub

Over to you – sample answer
Mentioned in B: central and peripheral cyanosis, clubbing, irregular pulse (in time and force), thrill, murmur, friction rub

24.1 1j, 2e, 3a, 4h, 5b, 6c, 7f, 8d, 9g, 10i

24.2 1 fever 2 rigors 3 incubation 4 prophylaxis 5 parasites

24.3 1 curable 2 outbreak 3 microorganisms 4 afebrile/apyrexial 5 glandular fever

25.1

Noun	Adjective
affect	affective
anxiety	anxious
behaviour	behavioural
dementia	demented
disturbance	disturbed
suicide	suicidal

25.2

1 behavioural syndrome
2 eating disorder
3 major depression
4 mental retardation
5 personality disorder
6 psychomotor retardation
7 sleep disturbance
8 substance abuse

25.3

1 behaviour 2 mood 3 panic attack 4 senile dementia 5 functional 6 compulsion

25.4

The patient has three of the listed symptoms:
tiredness (fatigue)
loss of interest
sleeping poorly (insomnia)

Over to you – sample answer
She does not meet the requirements for major depression as she has only three of the symptoms. (In fact, investigations revealed that she was suffering from hypercalcaemia due to primary hyperparathyroidism.)

26.1

Adjective	Noun
blind	blindness
conscious	consciousness
deaf	deafness
dizzy	dizziness
numb	numbness
light-headed	light-headedness
unsteady	unsteadiness

26.2

double vision
epileptic fit
prodromal symptom
syncopal attack
urinary incontinence
visual acuity

26.3

1 consciousness
2 warning
3 passed
4 stress
5 dizzy
6 light-headed
7 control
8 involuntary

Over to you – sample answer
According to the article, the smells are strawberry, smoke, soap, menthol, cloves, pineapple, natural gas, lilac, lemon and leather.

27.1

Noun	Adjective
absence	absent
diminution	diminished
flaccidity	flaccid
spasticity	spastic
wasting	wasted

27.2

1 muscle tone
2 muscle bulk and possibly involuntary movements
3 power
4 coordination

27.3 1 flaccid 2 Brisk 3 wasting 4 involuntary 5 Babinski/plantar 6 coordination 7 tendon/reflex hammer 8 diminished

27.4 GCS 5 (eye opening: to speech = 3, verbal response: none = 1, motor response: none = 1)

Over to you – sample answer
jaw jerk, biceps jerk, triceps jerk, supinator (wrist) jerk, knee jerk, ankle jerk
Diagnosis: the patient has Parkinson's disease.

28.1

Verb	Noun(s)	Adjective(s)
cure	cure	curative
excise	excision	
grow	growth	growing
invade	invasion, invasiveness	invasive
obstruct	obstruction	obstructive
palliate	palliation	palliative
palpate	palpation	palpable
spread	spread	spreading
swell	swelling	swelling, swollen

28.2 6, 3, 7, 2, 8, 4, 5, 1

28.3 1 spread 2 obstruction 3 painless 4 malignant 5 benign 6 Secondary 7 palliative 8 palpable

Over to you – sample answer
I'm afraid you have a condition called lymphoma, which is a tumour of certain white blood cells called lymphocytes. There are different types of lymphoma and we need to do more tests to find out which particular type you have. Some types require no immediate treatment. For others you may need drug treatment or radiotherapy.

29.1 1 premature 2 termination 3 trimester 4 labour, section 5 presentation 6 miscarriage 7 (umbilical) cord, stillborn

29.2

Verb	Noun
abort	abortion
deliver	delivery
induce	induction
miscarry	miscarriage
present	presentation
terminate	termination

29.3 1 terminated 2 aborted 3 delivered 4 induction 5 presented 6 delivered

30.1 blood-stained
breath sounds
pleural rub
productive cough
vocal resonance

30.2 1 Do you cough up any phlegm? / Is it a loose cough?
2 What colour is the phlegm?
3 Is it ever yellow?
4 Have you noticed any blood in it?
5 Any problems with your breathing?

30.3 1 true – A productive cough is often described as loose ... A cough may be productive, where the patient coughs up sputum (or phlegm)
2 false – crackles ... suggest the presence of fluid in the lungs
3 false – A cough may be ... non-productive, where there is no sputum
4 false – wheezes ... indicate narrowing of the airways. The sound of an asthma patient's breathing is also called wheeze
5 false – The sound heard when the pleural surfaces are inflamed, as in pleurisy, is called a pleural rub

Over to you – sample answer
A 36-year-old man complained of sudden right-sided chest pain with shortness of breath/ breathlessness/dyspnoea, while watching television. The pain was made worse by deep breaths and by coughing. The shortness of breath persisted over the 4 hours from its onset to his arrival in the Accident & Emergency department. He had a slight non-productive cough. There was no relevant past medical history or family history. He had had a three-week holiday in Australia three weeks previously.

On examination his temperature was 37.4°C, his respiratory rate was 24/min, his jugular venous pressure was raised 3 cm, his blood pressure was 110/64, and his pulse rate 128/min. In the respiratory system, expansion was reduced because of pain. There was a pleural rub over the right lower zone posteriorly. There were no other added sounds. Otherwise no abnormality was detected.

Diagnosis: This man had a pulmonary embolus.

31.1 1 Shingles 2 spots 3 small blisters 4 spots/blisters 5 small blisters 6 filled with pus 7 scabs 8 scars

31.2 location and distribution: first behind the ears and on the forehead then the trunk and limbs
grouping: scattered
type of lesion: macules
colour: pink

31.3 location and distribution: widespread on the chest and abdomen
grouping: scattered
type of lesion: small macules with some scales
colour: pink
(guttate psoriasis)

31.4 location and distribution: below lateral angle of the left eye
grouping: single
type of lesion: nodule
colour: white/pink
(basal cell carcinoma)

32.1

Common word	Medical term	Type of force
bruise	contusion, haematoma	blunt
cut	incised wound	sharp
graze	abrasion	blunt
scratch	linear abrasion	blunt
stab wound	penetrating wound	sharp
tear	laceration	blunt

32.2 1 grazes 2 contusion 3 tears

32.3 There is a laceration/tear on the left shoulder and an incised wound / a cut approximately 6 cm in length above the left nipple.

32.4 1 sore 2 stab/penetrating 3 blow/punch 4 superficial

33.1 1 frequency 2 nocturia 3 dysuria 4 hesitancy 5 stream/flow 6 dribbling

33.2 1b, 2a, 3e, 4f, 5d, 6g, 7c

33.3 Possible questions:

frequency: How often is that? How often do you pass urine?
dysuria: Do you get any burning or pain when you pass water?
nocturia: Do you have to get up at night?
hesitancy: Do you have any trouble getting started?
incontinence: Do you ever lose control of your bladder?
haematuria: Have you ever passed blood in the urine?
urgency: Do you have to rush/hurry to get to the toilet in time?

33.4 There was a trace of blood, gross proteinuria, and no casts.

Over to you – sample answer
Mr Jones has early prostatic hypertrophy.

34.1 1b, 2c, 3a, 4d

34.2 1 cuff 2 diaphragm 3 pump 4 valve, gauge 5 systolic blood pressure 6 deflate 7 diastolic

34.3 1 Phlebotomists 2 venipuncture 3 fist 4 tourniquet 5 dressing 6 bruising 7 specimen tubes 8 laboratory

35.1

Suspected condition	Test
anaemia	EDTA blood
bacterial conjunctivitis	Eye swab
genital herpes	Swab in Virus TM
meningitis	CSF
septicaemia	Blood culture
urinary infection	MSU
urinary infection (catheter in place)	CSU

35.2 1 within normal limits, grams 2 low, micromols per 3 elevated, units per litre 4 units per litre 5 normal, micromols per litre

35.3 Possible answer:
Sodium is normal, one hundred and thirty-eight millimols per litre.
Potassium is within normal limits, four point five millimols per litre.
White cell count is elevated, twelve point two times ten to the power nine per litre.
Haematocrit is low, zero point two two four litres per litre.
Mean corpuscular volume is down, seventy-two point five femtolitres.
Alkaline phosphatase is unremarkable, seventy-two units per litre.
Alanine aminotransferase is reduced, nine units per litre.

36.1

Verb	Noun
consent	consent
excise	excision
incise	incision
insert	insertion
recover	recovery
swallow	swallow

36.2 1 feed/insert/introduce 2 jelly 3 shaft/tube 4 cauterize 5 polyp 6 excise 7 biopsy 8 rigid 9 local anaesthetic 10 get used to

36.3 1 pulse oximeter 2 in the left lateral position 3 nasal cannula 4 premedication 5 introduced 6 transferred

37.1 1 medium 2 radiopaque 3 enema 4 serial 5 blurred 6 inflated

37.2 1 radiology 2 radiography 3 radiotherapy 4 radiopaque 5 radiolucent 6 radiographer

37.3 1 facing 2 Push 3 out 4 still 5 take 6 hold 7 sideways

37.4

Verb	Noun	Adjective
	abnormality	abnormal
breathe	breath	
drain	drainage	
intervene	intervention	interventional
	therapy	therapeutic

38.1 1c, 2d, 3a, 4b

38.2 a10, b6, c8, d4, e9, f5

38.3 a12, b17, c11, d13, e14

38.4 breathe in
excise diseased tissue
experience discomfort
foreign bodies
hold your breath
informed consent
introduce the endoscope
recovery area
local anaesthetic

1 local anaesthetic 3 recovery area 5 informed consent
2 foreign bodies 4 excise diseased tissue

39.1
1 tracing
2 exercise tolerance test
3 a skipped heart beat
4 electrolyte disturbance
5 conduction
6 screening
7 calibrate
8 stylus

39.2
a RA / right arm
b RL / right leg
c LL / left leg
d LA / left arm
e Chest positions

39.3 1 rate 2 complexes 3 wave 4 leads 5 interval

40.1 1e, 2g, 3i, 4b, 5a, 6c, 7d, 8f, 9h

40.2 1 Irritation 2 chemist's 3 side-effect 4 contraindicated 5 blistering 6 dose 7 Cautions 8 Indications 9 pharmacist 10 pharmacy

40.3 Streptokinase, one and a half million units by intravenous infusion over sixty minutes.
Aspirin, three hundred milligrams, by mouth, immediately.
Diamorphine, two point five to five milligrams, intravenously, immediately.
Metoclopramide, ten milligrams, intravenously, immediately.
GTN, three hundred micrograms per five millilitres, by intravenous infusion. Start at forty micrograms per minute.

41.1 1 scalpel 2 retractor 3 scissors 4 artery forceps

41.2
1 assistant
2 prepping/preparing
3 drapes
4 sterile
5 retractor
6 swabs/sucker
7 sucker/swabs
8 ligatures
9 drain
10 sutures/stitches/staples

41.3 1 divided 2 repaired 3 mobilized 4 excised 5 infiltration 6 preserving 7 staples 8 redundant 9 closure 10 layers

42.1 1 chemotherapy 2 radiotherapy 3 physiotherapy 4 Cognitive Behavioural Therapy 5 physiotherapy 6 Cognitive Behavioural Therapy

42.2 1 Curative 2 adjuvant 3 Palliative 4 implants 5 fraction

42.3 1 referrals 2 therapy 3 replacements 4 physiotherapist 5 Rehabilitating

43.1 1 Risk factors 2 false negatives 3 false positives 4 exposed to 5 booster 6 contracted 7 resistant to 8 outbreak

43.2 Possible answers:
1 Women aged from 50 to 70 should have mammography every three years to check for breast cancer.
2 Patients with heart disease should have a blood cholesterol test every six months to check their cholesterol level.
3 Women between 20 and 60 should have a smear test every three years to check for cervical cancer.
4 Patients over 40 with high risk factors should have their blood cholesterol checked every year.
5 Patients with diabetes should have ophthalmoscopy every year to check for diabetic retinopathy.
6 Pregnant women should have the AFP test between 16 and 17 weeks to check for neural tube defects and Down's Syndrome risk.

43.3
1 hepatitis A, malaria
2 hepatitis A, malaria, typhoid
3 hepatitis A, hepatitis B, malaria, typhoid
4 hepatitis A, rabies, Japanese encephalitis, malaria, typhoid
5 a booster dose for tetanus

44.1 1 case fatality 2 death/mortality 3 birth 4 survival 5 prevalence 6 incidence

44.2 1 more common 2 highest 3 common 4 lowest 5 low 6 uncommon/rare

44.3 affected by
association between
incidence of
lead to
rare in

1 incidence of 2 association between 3 lead to 4 rare in 5 affected by

45.1 a9, b7, c8, d3, e4, f2

45.2 1e, 2d, 3a, 4g, 5b, 6f, 7h, 8c

46.1

Noun	Verb
bias	bias
control	control
exposure	expose
participant (person)	participate
intervention	intervene
study	study

46.2

1 controls
2 double-blind
3 longitudinal
4 cohort
5 randomized
6 placebo
7 risk
8 confounding

46.3

1 longitudinal cohort study
2 double-blind, randomized controlled trial
3 case-control study
4 case-control study

47.1

1 Do you have a partner?
2 What do you do for a living?
3 Can you describe the pain?
4 Where does it hurt?
5 Does it go anywhere else?
6 When does it start?
7 How long does it last?
8 Does anything bring it on?
9 Does anything make it better?
10 Have you any other problems related to the pain?

47.2

1 Which part of your head is affected?
2 Could you describe the pain?
3 How long do they last?
4 Does anything bring them on?
5 Does anything make them better?
6 Is there anything else that affects them?
7 do you feel anything else wrong when it's there?

47.3

1 gnawing, burning
2 stinging, scalding
3 sharp, burning
4 crampy, colicky
5 throbbing
6 ache
7 sharp
8 gripping

48.1

1 over-the-counter
2 dose
3 allergy
4 herbal remedy
5 siblings
6 Recreation
7 Housing
8 compliance

48.2

1 Are your parents alive and well?
2 How old was your father when he died?
3 Do you know the cause of death? / What did he die of?
4 Do you have any brothers and sisters / siblings?
5 Are all your close relatives alive?
6 What did he die of?
7 Does anyone in your family have a serious illness?
8 (As far as you know,) is anyone taking regular medication?
9 Do you have any children?

48.3

1 What kind of house do you live in?
2 Are any of them at nursery or school?
3 Do you have any financial problems?
4 Do you have any hobbies or interests?
5 Do you smoke? How many a day?
6 Have you tried giving up?
7 How much do you drink in a week?
8 Can you give up alcohol when you want?

49.1 1a, 2c, 3e, 4c, 5b, 6d, 7f

49.2 1 I, 2 I, 3 C, 4 E

49.3 1 come on 2 put on 3 give up 4 carry on 5 bring up 6 turned out

Over to you – sample answer

1 Have you had any pain or problems with your mouth?
2 Is it difficult for you to swallow food or drink?
3 Have you had any discomfort after eating?
4 Do you ever get a burning feeling in your chest?
5 Do you have any pains in your stomach?
6 Have you lost any weight?
7 Have you noticed any change in your bowel habit?
8 What colour are your bowel movements?
9 Have you noticed any blood in your stool?
10 Does this come mixed with the stool or before or after?

50.1 1 Look, touch 2 Sit, let 3 Close 4 Take, off 5 Touch 6 Turn 7 Lie 8 Keep 9 Roll up 10 Look at

50.2

1 Touch your shoulders with your hands.
2 Put your hands behind your head.
3 Put your hands behind your back.
4 Raise your arms above your head.
5 Bend your head forward … backward.
6 Bend your head to the right … to the left.
7 Turn your head to the right … to the left.
8 Bend backwards.
9 Touch your toes.
10 With your heel on the ground, turn your foot as far as you can.
11 Bend your knee.
12 Bend to the left … to the right.
13 Bend your toes up and down.

50.3

1 do you know what we're going to do this morning
2 what we do is
3 a little bit of discomfort
4 take very long
5 over
6 could you (just)
7 I'm going to
8 ready
9 you'll feel
10 very well
11 over

51.1 1 delusion 2 hallucinations 3 illusion 4 obsessional 5 deluded 6 disorientation

51.2

Noun	Adjective
confusion	confused
delusion	deluded
depression	depressive (illness)
	depressed (patient)
disorientation	disoriented
obsession	obsessional (symptoms, thoughts)
	obsessive compulsive (disorder)
psychiatry (field), psychiatrist (practitioner)	psychiatric

1 psychiatric 2 depression 3 depressed 4 disoriented

51.3

1 Can you describe your mood at the moment?
2 Do you take pleasure in anything?
3 How are your energy levels?
4 How long have you been feeling like this?
5 How are you sleeping?
6 What's your appetite like?
7 Have you noticed any change in your weight?
8 Can you keep your mind on things?
9 What do you feel the future holds for you?
10 Have you ever thought of suicide?

52.1

1 c
2 a, f
3 d
4 b, e

52.2

1 a stroke
2 tube which carries urine from the bladder
3 needing to pass urine frequently at night
4 breathlessness
5 a painkiller, medicine to reduce swelling
6 tablets to improve your mood

52.3

1 developed
2 (mainly) because you are
3 This is why / It's the reason why
4 give you advice on
5 make you an appointment with / arrange for you to see
6 going to start you on
7 should / should try to
8 want you to
9 arrange for you
10 Hopefully we can
11 anything you'd like to ask

53.1 1 carry on 2 gave up 3 end up 4 start, off 5 Cut down 6 settle 7 avoid

53.2 1c, 2f, 3e, 4a, 5b, 6d

53.3 Possible answers:
You should keep to a low cholesterol diet.
You should keep up any regular physical activity you are used to.
You shouldn't do any activity that brings on angina.
You should avoid moving from floor to standing exercises too quickly.

Over to you – sample answer
The first option is to do nothing. The fibroid will shrink when you become menopausal, although we can't be sure when that will be.

If you prefer, I can refer you to a gynaecologist to discuss surgical treatment. There are three possibilities: one is embolization which is relatively minor. It means closing off the artery that feeds the fibroid. The second would be removal of the fibroid alone without removing your womb. This needs abdominal surgery, as does a hysterectomy when your womb is removed.

You don't have to decide today. I can give you leaflets and recommend websites for you to look at. Then come back and see me in two weeks' time.

54.1
1 I'd like to record this consultation
2 I'm afraid / I'm sorry to have to tell you
3 isn't an option
4 a lot we can do to help you
5 make you more comfortable
6 never be certain about these things
7 it's a matter of months rather than years
8 I'm sorry to have to tell you
9 book you into
10 Could you tell me what

54.2 1 with 2 to 3 with 4 with 5 into 6 for

Over to you – sample answer
There are two questions about tumours in the brain – 'Is it dangerous?' and 'Can it be removed?' We know that yours isn't cancerous. Can we remove it? Well, it's quite large but I believe that if we can get as much of it out as possible, there's a good chance we can cure you. If we do nothing, then I'm afraid it could kill you – not in the short term but perhaps in five years' time. An operation like this on the brain carries significant risk. You could be disabled, or your personality could be affected. So you need to think carefully about whether or not you want to go ahead with the operation.

55.1 1 shows 2 compared 3 threefold 4 doubled 5 trebled 6 twofold

55.2 1 less 2 almost 3 under 4 over 5 about 6 around 7 less 8 approximately 9 nearly

56.1

Verb	Noun
de'crease	'decrease
drop	drop
fall	fall
in'crease	'increase
rise	rise

56.2 1 dropped 2 rose 3 gradually 4 increased 5 steadily 6 fell 7 reaching a peak 8 accounted

56.3 1 line graph / bar chart 2 bar chart 3 pie chart

57.1
1 Results
2 Introduction (Objective of the research)
3 Method (Statistical analysis)
4 Method (Subjects)
5 Discussion (Main finding)
6 Introduction (Background)
7 Discussion (Limitation)
8 Results

57.2 Possible answers:
1 We assessed whether calcium and vitamin D supplementation reduce the risk of fractures in postmenopausal women.
2 The aim of our study was to determine whether the way doctors dress influences patients' confidence and trust in them.
3 This study evaluated the risk of HHV–8 transmission by blood transfusion.
4 We investigated the association between never being married and increased risk of death.

57.3 Possible answers:
1 This study failed to show that calcium and vitamin D supplementation reduce the risk of fractures in postmenopausal women.
2 We have shown that the way doctors dress influences patients' confidence and trust in them.
3 This study provides strong evidence of HHV–8 transmission by blood transfusion.
4 These results suggest that there is an association between never being married and increased risk of death.

58.1 1 objective 2 setting 3 subjects 4 outcome 5 design

58.2
1 Past employees of Shell Oil who retired at ages 55, 60, and 65 between 1 January 1973 and 31 December 2003
2 To assess whether early retirement is associated with better survival
3 Petroleum and petrochemical industry, United States
4 Hazard ratio of death adjusted for sex, year of entry to study, and socioeconomic status
5 Long term prospective cohort study
6 No

58.3 The correct order is: 3 (Objective), 7 (Design), 1 (Setting), 6 (Subjects), 2 (Main outcome measure), 4 (Results), 5 (Conclusion)

58.4 1d, 2b, 3a, 4e, 5c

59.1 The correct order is: 4 (Topic), 9 (Pathophysiology), 5 (Sources), 6 (Sources), 2 (Diagnosis), 8 (Diagnosis), 3 (Diagnosis), 10 (Diagnosis), 7 (Treatment), 1 (Treatment)

59.2

Emphasizing	I think it's important to emphasize that …
Listing	First of all …
Exemplifying	such as …
Contrasting	however …
Summing up	So …
Changing topic	How do we diagnose it? Now, in relation to …
Referring to a slide	On the slide here you'll see …
Announcing the topic	I'd like to tell you about …

60.1

c/o	complained of
2/52	2 weeks
PH	past history
FH	family history
MI	myocardial infarction
BP	blood pressure
1/12	one month
nil	nothing
SH	social history
a&w	alive and well
OE	on examination
CXR	chest X-ray

60.2 The correct order is: 1, 5, 8, 9, 2, 3, 4, 6, 7
(This patient had carcinoma of the bronchus.)

60.3 Mr McNamara, 63, taxi driver
c/o shortness of breath 3/12
ankle swelling 3/12
chronic cough, purulent sputum, occasional haemoptysis
PH partial gastrectomy, 1980

OE pale T 37°C
leg oedema
no clubbing or lymphadenopathy
chest NAD
liver 5cm palpable smooth and non-tender
scar of previous operation

(Mr McNamara had pulmonary tuberculosis.)

Index